W9-BIS-608

7/98

ALIEN LIFE

The Search for Extraterrestrials and Beyond

ALIEN LIFE
The Search for Extraterrestrials and Beyond

BARRY PARKER

Drawings by **Lori Scoffield**

PLENUM TRADE • NEW YORK AND LONDON

Library of Congress Cataloging-in-Publication Data

Parker, Barry R.
    Alien life : the search for extraterrestrials and beyond / Barry
Parker.
        p.   cm.
    Includes bibliographical references and index.
    ISBN 0-306-45795-4
    1. Life on other planets. 2. Extraterrestrial anthropology.
I. Title.
QB54.P33 1998
576.8'39--dc21   98-4552
                                                              CIP

ISBN 0-306-45795-4

© 1998 Barry Parker
Plenum Press is a Division of Plenum Publishing Corporation
233 Spring Street, New York, N.Y. 10013-1578

http://www.plenum.com

Printed in the United States of America

Other Recommended Books by Barry Parker

Chaos in the Cosmos
The Stunning Complexity of the Universe

Stairway to the Stars
The Story of the World's Largest Observatory

The Vindication of the Big Bang
Breakthroughs and Barriers

Cosmic Time Travel
A Scientific Odyssey

Colliding Galaxies
The Universe in Turmoil

Invisible Matter and the Fate of the Universe

Creation
The Story of the Origin and Evolution of the Universe

Search for a Supertheory
From Atoms to Superstrings

Einstein's Dream
The Search for a Unified Theory of the Universe

Contents

Preface

"A grotesque-looking gray creature with leathery skin emerged from the craft. It was hairless, with a large head and dark, unfocused eyes. Its mouth was small and lipless, and its body quivered and pulsed as it made its way down the steps. When it reached the bottom it stared at us for several seconds, then raised one of its strange appendages in what appeared to be a greeting."

The above sounds like something from a science fiction novel but as difficult as it is to believe, it may one day be part of a news story. Alien life almost certainly exists. We have already found striking evidence for microbes on Mars, and the moons Europa and Titan have conditions that are conducive to elementary forms of life. Furthermore, beyond our solar system are billions of stars, many with planets that may harbor life. We have, in fact, recently discovered planets around several nearby stars, and although none appear to harbor life, they are an encouraging sign. As our technology advances, many more extrasolar planets will no doubt be found, and it is only a matter of time until one of them shows signs of life. The first forms will likely be elementary but eventually we may detect civilizations—perhaps far more advanced than us.

Unanswered questions about aliens abound. What would they look like? What would the first meeting with aliens be like? Would they be menacing or friendly? And what about the strange objects—the UFOs—that are reported almost daily around the world? Are they aliens? Furthermore, many people have claimed to have been abducted by aliens and taken into flying saucers and examined. Are these cases being taken seriously?

In this book I have tried to answer many of these questions, and I have tried to give you a scientific account of what astronomers and other scientists are doing in their search for extraterrestrial life. Considerable controversy surrounds the study, many of the ideas are speculative and far-out, and some ideas will no doubt turn out to be wrong, but speculation is always interesting and frequently helpful, and some has been included.

Technical terms have been kept to a minimum, but there may be words you are unfamiliar with. Because of this I have included a glossary.

I am grateful to the people who have helped me with this book. They include Everett Gibson, Jr., Nadine Barlow, Randy Spaulding, and Rajat Kudchadker.

The line drawings, sketches, and paintings were done by Lori Scoffield. I would like to thank her for an excellent job. I would also like to thank Linda Greenspan Regan, her assistant, Vanessa Tibbits, and the staff of Plenum for their assistance in bringing the book to its final form. Finally, I would like to thank my wife for her support while the book was being written.

chapter 1

Introduction

Looking into the sky on a clear dark night, we are struck by the beauty and grandeur of the stars. There appear to be tens of thousands of them, but in reality even on the clearest nights we see only about 2000. This is, however, only a tiny fraction of the number that is actually there. Our universe is immense beyond imagination, so vast that a light beam traveling at 186,000 miles per second would take 15 billion years to cross it.

As we look at the stars, we see a tremendous panorama: stars of all colors, giant stars, and average stars like our sun. Some are dim, some bright; some sparkle, others shine steadily. And stretching across the sky from horizon to horizon is a faint ribbon of light. Through a pair of binoculars this stream is resolved into thousands of stars. This is our view of the system we live in—the Milky Way galaxy. Containing 200 billion stars, it is only one of billions of galaxies in the universe. There may, indeed, be as many galaxies in the universe as there are stars in our galaxy.

As we stare at the stars our imagination stirs. Is it possible that someone on a planet near one of these stars is also looking into his night sky, asking the same questions we are asking, wondering if there is life beyond his planet? Is it possible that one of these stars has a planet whirling around it, teaming with life, as Earth is? The discovery of life elsewhere in the universe would be the greatest discovery ever made by humankind. The implications would excite everyone from the smallest child to the oldest adult, and indeed, many astronomers say it is only a matter of time before it happens.

There are over a million stars in this photograph. The probability that at least one of them has a planet that harbors life seems high. (Courtesy Lick Observatory)

Scientists believe that life on Earth began as an evolutionary process. The right atmosphere, the right conditions, and the right energy sources were all available, and they came together in exactly the right way. There is no doubt that we are special. It took a lot of special conditions to produce us, and if the slate were ever wiped clean, in other words, if the Earth's inhabitants were wiped out, whatever emerged out of the ashes (if indeed anything did) would not resemble the humans that now populate our planet. Still, if humans evolved once, even with the odds being small—extremely small—it could happen again on some distant planet orbiting a star similar to our sun. All that is needed are the proper conditions and the right molecules.

Through our telescopes we see billions of stars, and from recent indications many, perhaps most, have planets—even systems of planets—as our sun does. That would amount to a large number of planets. Is it possible that they are all barren of life? It doesn't seem likely, yet strangely we haven't found a single sign of life beyond our solar system. Recently, however, we have found indications that lower forms of life—microbes—may exist elsewhere within our solar system. One of the great discoveries of 1996 was a meteorite from Mars that appeared to contain microbes. If this finding is validated, we would finally have proof that there is life beyond Earth, even if it is an elementary form.

And Mars may not be the only abode for life (aside from Earth) in the solar system. The moons Europa and Titan have both recently attracted a great deal of attention. Close-ups of Europa by the satellite *Galileo*, along with other data, indicate that it is covered with a layer of ice, with liquid water beneath the ice—water that may be loaded with microbes and hydrocarbons. And beneath the dense atmosphere of Titan may lie gigantic lakes, crawling with elementary forms of life.

Some of the most exciting news in recent months is the discovery of planets beyond our solar system—extrasolar planets. Over a dozen are now known and NASA has plans for launching a telescope into space that may allow us to find hundreds more. So far, all are Jupiter-sized and none likely to contain a higher form of life, but lower forms are possible. As our technology advances, however, smaller planets like the Earth will no doubt be found, and there is a much better chance that some of them will contain life.

Our best bet for detecting life, however, is the large radio telescopes that are now being directed toward the stars; they are searching for signals from an advanced civilization—a race of intelligent beings with a technol-

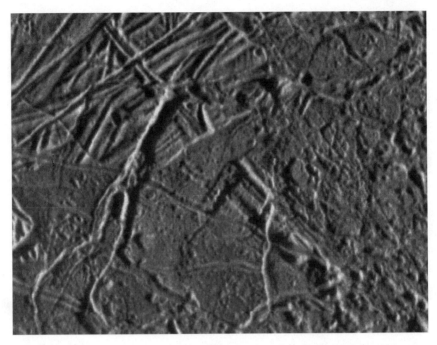

A close-up of the icy surface of Europa. There may be life in the ocean below. (Courtesy NASA)

ogy at least equivalent to ours. With the latest instrumentation, millions of different frequency channels can be listened to at once, and within a few years the thousand nearest stars will be thoroughly checked. The chances appear high that at least one of them will have some form of life.

Even if the life is only an elementary form, we could still detect it. The spectrum of a planet would give us considerable information. If its spectrum indicated water, oxygen (ozone), and carbon dioxide, it would be a good candidate. Within a few years we will be able to get the spectrum of planets orbiting nearby stars, and if there is any indication that these molecules are present, our radio telescopes will be quickly trained on them to see if there are any signs of life.

Although the chances seem overwhelming that there is life out there somewhere, astronomers have uncovered an enigma that could have serious implications. As strange as it may seem, several astronomers have pointed out that we may be the only advanced civilization in our galaxy. The reason for their belief: our galaxy has been around for 16 billion years,

Radio telescopes of the Large Array in New Mexico. Radio telescopes are our best bet for locating extraterrestrial civilizations. (Courtesy National Radio Astronomy Observatory)

The 1000-foot disk of the National Astronomy and Ionosphere Center near Arecibo, Puerto Rico. It is the largest radio telescope in the world. (Courtesy National Astronomy and Ionosphere Center)

and if civilizations have emerged, many of them would have emerged millions of years ago, perhaps even billions of years ago. It is difficult to imagine what a civilization a million years old would have accomplished, considering what we have accomplished in only a few centuries. It certainly would have the ability to visit nearby stars. We have traveled to the moon and beyond, and we now have spaceships headed to the stars. If such a civilization were sufficently advanced, it would no doubt spread out and colonize the stars close to it, and these new colonies would, in turn, send out their own spaceships for further colonization. In fact, it is easy to show that a civilization a million years old would have colonized much of our galaxy by now. There should therefore be millions of civilizations out there, and if there are, some of them should have visited Earth.

It is possible, however, that we *are* the only civilization in our galaxy. The tremendous distances between stars, the tremendous amount of fuel needed for such trips and so on, are problems that even supercivilizations may not have overcome. Millions of civilizations may have arisen, enjoyed a high standard of living for a few centuries or perhaps several thousand years, and then, as their resources dwindled, died off.

It is extremely difficult using known technology—even energy sources and ideas that are well beyond us—to get to the stars. Even with extremely fast spaceships it would take decades to get to the nearest stars, and it would be impossible to travel freely across large sections of our galaxy. With speeds very close to that of light it would take over 50,000 years to get halfway across our galaxy. Only civilizations that have discovered exotic methods of travel—traversable wormholes through space, the use of another dimension, or the use of warp speeds much greater than that of light—could truly become stellar explorers.

Many people are convinced that extraterrestrials are already here and that we are seeing them almost daily around the Earth in the form of UFOs. Astronomers (and scientists in general) are not convinced this is the case. There are, indeed, a large number of interesting UFO cases, some that seem unexplainable. Unfortunately, we have no really hard evidence that any of the objects seen are extraterrestrial. There have been no captured spaceships, no captured aliens, and nothing left by the aliens that would indicate they were from another world and had a highly advanced technology.

Some of the most interesting UFO cases are the recent abductee cases, where the witnesses claim that they were transported into a spaceship and examined by humanoid creatures. Because of the large number of these cases lately, scientists have begun to take an interest in them. Most scientists are at a loss, however, as to how to explain them.

In the pages that follow we will look at the possibility that there is indeed alien life out there, and the implication of such a discovery. In order to understand the details, however, we will have to begin with the basis of life—the basic molecules that make up life.

chapter 2

The Architecture
of Life

If we are to consider the possibility that life may exist elsewhere in the universe we must begin with life as we know it—life here on Earth. How did it come about? Was it a natural process? To answer questions like these we have to go to the basis, or essence of life. Everyone knows that living matter is different from inanimate matter, so different in fact that many early scientists thought that it was imbued with some sort of magic, as if there were a mysterious "life force" within the matter of life.

The matter of life is made up of the same type of atoms that make up all elements of nature. But, as you might expect, a small number predominate. Of the 85 known stable elements only 4 are involved in 95 percent of the atoms in living matter: nitrogen, oxygen, carbon, and hydrogen. They are, in a sense, the basic units of life, but for life as we know it they must come together as molecules, and they do this by forming bonds.

Atoms are made up of a nucleus of protons and neutrons, surrounded by a cloud of electrons. The electrons orbit the nucleus in shells, with two allowed in the first shell, eight in the second, and so on. In many atoms the shells are not full, and electrons can be shared between two atoms. One of the outer electrons of carbon, for example, can share an electron with a nearby hydrogen atom, like two skaters sharing a scarf by holding onto the two ends as they glide over the ice. When electrons are shared in this way a bond is formed between the two atoms.

Carbon is particularly adept at forming bonds. It has four electrons in its second shell and therefore four spaces where it can form bonds. Because

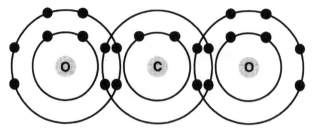

The bonding between carbon and two oxygen atoms. Bonding occurs because of a sharing of electrons.

of this it is able to produce long, stable chains of molecules of the type needed for life. Life on Earth is possible because of this; it is, in fact, based on carbon.

What Is Life?

Before we look into the basic molecules of life, let's consider life itself. What exactly is it? How do we define it? It might seem that we could define it in terms of change. As everyone knows, all life forms are in a continual state of change: cells die and are replaced by new ones. The cells of your body, for example, are generally different from the ones you had only eight years ago; most of them have renewed themselves. But your personality and physical appearance likely haven't changed significantly, so it's obvious that there must be a master plan somewhere in your body that allows duplication.

It's easy to see, however, that any attempt to define life in terms of change encounters difficulties. Many things, such as a stream or a flame, change continuously, and they are not alive. Perhaps a better way to define life would be in terms of reproduction. All life reproduces itself. But even here we have to be careful: crystals duplicate themselves, as do viruses, and we know that crystals are not alive. Whether viruses are alive or not is difficult to say; they seem to be very close to the border between the living and nonliving.

Reproduction, nevertheless, is the key to life. In the past few decades tremendous breakthroughs have been made in our understanding of the life process. The basic unit, the cell, has of course been known since the invention of the microscope, but as scientists have probed deeper and deeper into the cell, they have made some fascinating discoveries. Just as the atom has a nucleus, so too has the cell, and within it are units that control the growth and development of the cell. The controlling mechanism is in the form of three giant molecules: deoxyribonucleic acid (DNA), ribonucleic acid (RNA), and protein. One of the main concerns of early molecular biologists was, which of these three is fundamental? In other words, which is responsible for directing the replication and metabolism of the cell?

Protein or Nucleic Acid?

Early studies showed that protein was considerably more complex than either of the nucleic acids, and it was assumed that the "life essence," or more specifically, the "code of life" was contained in it. In fact, for years hardly anyone paid any attention to DNA and RNA, even though it was well-known that they played a role in the life process. They weren't

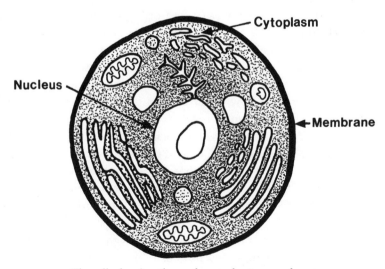

The cell, showing the nucleus and outer membrane.

considered worthy of detailed attention because they appeared to be too simple.

In the late 1930s and early 1940s, however, a series of experiments was performed that, over a period of years, led to a revolution in molecular biology. Ostwald Avery and two assistants, Colin MacLeod and Maclyn McCarty, of the Rockefeller Institute in New York, showed that protein wasn't the "master" molecule; it didn't contain the hereditary material of life. DNA was the central molecule—the Cinderella compound.

Avery's work centered on pneumonia-causing bacteria. Pneumonia was a fatal disease at the time, snuffing out thousands of lives each year, and Avery was sure he could make a contribution to the understanding of the problem. He had joined the Rockefeller Institute a few years after obtaining his medical degree in 1904. About this time something strange had just been discovered about pneumonia bacteria. Two strains called the R and S strains were known, and scientists had shown that living R bacteria could be added to dead S bacteria and bring it back to life. Both the R and S bacteria were known to be made up of protein and nucleic acid (DNA). Avery decided to find out which of these molecules was responsible for the transformation. Over a period of years he and his colleagues proved that it was DNA. Protein could be excluded in the experiment; it had no effect. When he published his discovery, it was a tremendous surprise to many.

The next step, of course, was to find out how DNA performed this miracle. Unfortunately, little was known about DNA at the time, other than that it was a long chain molecule—a polymer—made up of sugar and phosphate. Base units called purines and pyrimidines were also present, but it was not known how they were attached to the polymer. It was soon obvious that the secret of how the DNA brought the bacteria back to life lay in its structure or makeup.

Avery's paper caused a stir among scientists, but worldwide only a few were working on nucleic acid. One of these few was Edwin Chargaff of Columbia University. Born in Austria, Chargaff obtained a Ph.D. from the University of Vienna in 1928; he came to Columbia University in 1935.

Chargaff's main interest was the four base units in DNA: adenine, thymine, guanine, and cytosine. (They are usually designated by their first letters: A, T, G, and C.) He discovered that in a given sample of DNA they were not present in equal numbers, but the total amount of purine in DNA $(G + A)$ was always equal to the total amount of pyrimidine $(C + T)$.

Looking closer he found that the amount of A was always equal to the amount of T, and the amount of G was always equal to the amount of C. This would eventually turn out to be a vital clue to the structure of DNA.

Schrödinger

In the same year that Avery and his colleagues published their classic paper on DNA a well-known physicist, Erwin Schrödinger, who was living in Ireland, published a tiny book entitled *What Is Life?*

Born in Austria in 1887, Schrödinger obtained a Ph.D. from the University of Vienna in 1910. In 1926 he formulated quantum mechanics, a theory that revolutionized physical science. He was awarded the Nobel Prize for it in 1933. Unlike other great scientists such as Einstein, Heisenberg, and Pauli, who made their major contributions when they were in their twenties, Schrödinger formulated quantum mechanics when he was forty. Interestingly, a few years earlier he had almost abandoned physics for philosophy.

He fled Germany after Hitler came to power, settling finally in Ireland at the School for Advanced Studies in Dublin. His small book on life was published in 1944.

It might seem strange that a physicist would cross disciplines, but in later life Schrödinger was curious about many aspects of nature, and life was one of the great mysteries of the time.

In his little book he put forward the idea that somewhere within the basic molecule of life—at that time it was still considered to be protein—there was an "aperiodic crystal" that contained the code of life. Schrödinger assumed it was associated with the chromosomal fiber within the cells, since it was known that the chromosomes played an important role in mitosis. His aperiodic crystal was different from the usual periodic crystals of nature that were built of similar units. No "information" can be stored in a periodic crystal because all units are identical, but if the crystal were made up of a number of different basic units—the number need not be large—it could contain a tremendous amount of information (much as the two simple units 0 and 1 give an entire system of numbers—the binary system). In fact, the "code of life" could be contained in it.

Schrödinger's book was wrong in many respects; some of it, in fact, was known to be wrong at the time that he wrote it, and much has been proved wrong since. But the book was a delight to read. Schrödinger was not only a mathematical genius but also a superb writer, and he had a lot of insight. The book had a profound influence on many people, and as we will see, most of the people who made the breakthrough to the structure of DNA had read it.

Crick and Watson

Two of the major players in the drama that led to the discovery of the structure of DNA were Francis Crick and Jim Watson. The older of the two, Crick, was born in Northampton, England, in 1916. He went to University College in London, graduating in 1938 with a degree in physics. He started on his doctorate the following year, but World War II intervened and for the next several years he worked on radar and other war-related projects. When the war ended he had second thoughts about returning to physics.

During the war years molecular biology had begun to flourish. Crick quickly realized that it was a field of considerable promise and one that he, with his knowledge of physics and math, could make a contribution to. He had read Schrödinger's book and it inspired him to return to Cambridge. He stayed there for two years studying the nature of magnetic particles in cells, then in 1949 he joined the molecular biology group at Cavendish, where he began working on a doctoral thesis on the X-ray diffraction of protein. It was headed by the molecular biologist Max Perutz.

Talkative, occasionally overbearing according to some, and brilliant, Crick had a knack for coming up with innovative new ideas. He had taken an interest in DNA after hearing a lecture by Linus Pauling in 1946, but with a thesis to complete he had little time to think about DNA. Nevertheless, he was curious about it and realized it was important.

Watson was quite different from Crick, and over ten years younger, but with the same spirit of determination. He was a child prodigy who entered college at 15, graduated at 19, and had a Ph.D. at 22. He obtained his Ph.D. at the University of Indiana, working under the well-known microbiologist, Salvador Luria.

Upon completion of his doctorate, Watson headed for Copenhagen for graduate work, but like Crick he had also been inspired by Schrödinger's book and was sure that the big breakthroughs in biology over the next few years were going to come in molecular biology. In May 1951 he attended a talk by Maurice Wilkins on X-ray studies of DNA that reaf-

Jim Watson.

firmed his decision to switch to molecular biology, and with his future a little uncertain (his scholarship at Copenhagen had to be shifted and there were problems), he headed for Cambridge, England, where he joined Perutz's group. He was assigned a desk in the same room as Francis Crick.

Crick's and Watson's personalities and knowledge meshed well. Crick had a good knowledge of X-ray diffraction, physics, and mathematical methods; Watson had a strong background in biology, was creative, and had a good imagination. Although neither was an expert on all aspects of molecular biology, they were able to bounce ideas off each other, and within a short time of getting together they decided to tackle the structure of DNA. Furthermore, they knew how they would go about it. Pauling had just determined the structure of protein by building scale models of the molecule. They would do the same for DNA. They didn't have the detailed knowledge of chemistry or of chemical bonding that Pauling did, but that didn't deter them.

Ideas and imagination were important, but they needed more than good ideas. They needed the latest experimental data, in particular the latest X-ray diffraction photographs of DNA. Fortunately, Maurice Wilkins was only a short distance away at King's College in London, and he was

Francis Crick.

continuing his work on DNA with Raymond Gosling. Like Crick, Wilkins was a transplanted physicist. He had obtained a Ph.D. in physics before the war, and during the war had worked on radar and the Manhattan Project. The wartime uses of physics—in particular, the devastation brought by the atomic bomb—eventually soured him on physics, however, and after the war he turned to biology.

Wilkins had obtained some excellent X-ray plates of DNA but realized that he and Gosling were limited in their ability to interpret the patterns. Wilkins was convinced the patterns indicated a helical structure, likely a single helix, but beyond that he was at a loss. Somebody with more expertise in the area was needed, so they hired Rosalind Franklin.

Born in 1920, Franklin graduated from Cambridge in 1941; in 1952 she was in Paris working on the X-ray diffraction of coal and carbon compounds. When she was offered the position at King's College in London she decided it was time to return home. It looked like an excellent opportunity and she was eager to join the group. But to her disappointment, the collaboration turned into a disaster. Franklin understood that she was to take over all X-ray diffraction studies of DNA, but when she got to London she found that Wilkins was not ready to relinquish his interests in the project. He knew, as Crick and Watson did, that the structure of DNA was one of the most important problems in biology, and considerable honor would come to the person who determined it.

Further resentment resulted when Franklin discovered that Wilkins considered her a junior partner rather than a collaborator. Within a short time they were not talking to each other, and finally, to his dismay, Wilkins found that Franklin was keeping her work secret from him.

Despite the friction, Franklin gave the project everything she had. Working with Gosling, she set up a new X-ray diffraction apparatus and tested it thoroughly. It was considerably better than the war surplus equipment Wilkins had used earlier and soon gave excellent results. After taking her first photographs of DNA she decided to give a colloquium.

Crick and Watson had been toying with the structure of DNA with little success. They needed access to Wilkins's and Franklin's data, so when they heard that Franklin was going to give a colloquium they decided that one of them should attend. Watson elected to go, but at this stage he still did not have a good understanding of X-ray diffraction, and he understood very little of the lecture. Furthermore, to Crick's dismay, he took no notes.

On the basis of what little Watson understood, however, he and Crick decided to build a preliminary model. It was obvious to them that the molecule had to be helical, but they had no idea how many helical strands were involved. It could be single-, double-, or triple-stranded. They decided on a triple-stranded model. The backbone of the model was known to be made up of alternating units of phosphate and sugar (deoxyribose sugar); the four bases were attached to this chain.

Franklin had suggested in her colloquium that the backbone was on the outside, but Watson, in his ignorance, had missed this. He and Crick decided to put the backbone down the center. They built a scale model of their design and reported their success to Wilkins. A few days later Wilkins, along with Franklin and Gosling, came to Cambridge to see it.

Franklin saw immediately that the model was inconsistent with the data she had obtained, and she didn't hesitate to tell them. It was, in fact, deficient in several other respects. It told them nothing about how the genetic code was contained in the molecule, or how the molecule used it to replicate. In short, it was a failure, and it didn't take Crick and Watson long to realize it.

It was a blow to their egos. Still, although they were down, they weren't out, and within a short time they tried a new approach. At this stage they had spent little time thinking about replication. The importance of it, however, was impressed on Crick after a talk with a young mathematician, John Griffith. They had been to a colloquium on the steady state universe in which Thomas Gold had introduced the "perfect cosmological principle." Crick suggested that there might be a similar principle in molecular biology, and it seemed natural to associate it with self-replication. This led to a discussion about how the molecule could replicate. It seemed that the four bases A, T, G, and C would have to be intricately involved. Griffith mentioned that he would look into the possibility of pairing between them. A few days later he told Crick that, according to his calculations, adenine was attracted to thymine, and guanine to cytosine. Crick realized that this might be important, but it seems that he later forgot it. It would explain Chargaff's rules (the amount of A = T and the amount of G = C), but Crick hadn't heard of them. Watson, on the other hand, was familiar with Chargaff's rules but hadn't mentioned them to Crick, or at least hadn't impressed their importance on him.

At this stage Crick and Watson weren't in a rush. Franklin had told them she was sure the structure wasn't a helix, and they were sure it was. At the end of 1952, however, their nonchalance came to an abrupt end. One of the desks in their room had been assigned to Peter Pauling (son of Linus Pauling), who had come to Cambridge for graduate work. In January 1953 he received a letter from his father telling him that he (Linus) and an associate, Robert Corey, had determined the structure of DNA. Linus would send him a preprint as soon as it was ready.

Crick and Watson were shocked. Linus Pauling had worked out the structure of protein, and they had known it was only a matter of time before he turned his attention to DNA. They worried that all of their work was down the drain. They had been "scooped." But when the paper arrived they were surprised. Pauling was suggesting a triple helix, quite similar to the one they had devised earlier. As they studied the details they saw that it was flawed: Pauling had made the same mistake as they had earlier. They were sure, however, that once the paper was published it wouldn't be long before someone pointed out the error to him, if indeed he didn't find it himself. They realized that at best they had six weeks to complete their model. They set to work immediately.

Watson was anxious to show Pauling's paper to Wilkins, and within a few days he was off to London. When he arrived at King's he was unable to find Wilkins, so he wandered into Franklin's lab and began telling her about Pauling's triple helix. Watson's description of this event in his popular book *The Double Helix* is one of the more controversial parts of the book. Many people feel he was unfair in his characterization of Franklin. According to his description, she flew into a rage over his insistence that the structure was helical and he thought she was going to attack him. The book failed to mention, of course, that he was over six feet tall and she was slim and barely over five feet.

Anyway, Watson retreated to the hall where he bumped into Wilkins. After he described "the attack" to Wilkins, a new camaraderie developed between the two men. Wilkins took Watson back to his office where he gave him one of the best DNA diffraction plates that had been taken by Franklin. The DNA was in a different form, which was referred to as the B form to distinguish it from the A form, and the patterns were much clearer.

According to Watson, his mouth dropped when he saw it. He realized that there was no doubt now that the structure was helical; furthermore, this pattern was much simpler than the A pattern, and easier to interpret.

He showed the photograph to Crick, and they quickly decided that their future efforts would be directed toward this new form. They were anxious to start building models again, and within a few days they had ordered scaled components of the molecule from the machine shop. They would build the molecule—piece it together like a jigsaw puzzle. But strangely, when the pieces arrived they stubbornly stuck to a model with the bases on the outside and the sugar–phosphate chain in the center. Finally, almost reluctantly they began switching things around, putting the bases on the inside and the phosphate–sugar spine on the outside. But there was still a major problem. How did the bases fit together? They tried fitting A with A, T with T, and so on, but nothing seemed to work.

Then fate stepped in. Their major difficulty was getting the bases to fit together, and the reason for this was that they were assuming the wrong form for them. Bases such as A, T, G, and C come in two slightly different forms called keto and enol. Crick and Watson had assumed they were in the keto form. This is what all the textbooks had told them, and they never questioned it. But a visitor to Cambridge, an American named Jerry Donahue, knew better. He was an expert on bonding in molecules, in particular, hydrogen bonding, and he knew that the textbooks were wrong. In a casual conversation he mentioned it to Crick. Crick found it hard to believe at first, but when he took the information back to Watson and they incorporated it into their cardboard model the pieces of the puzzle came together. The enol form was much more amenable to hydrogen bonding, and A now matched T across a hydrogen bond, and G matched C. This was Chargaff's rules and it made sense.

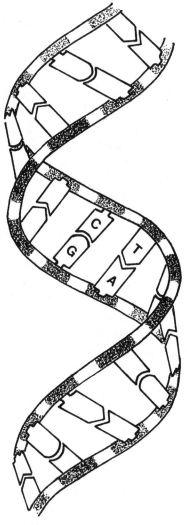

The DNA molecule showing its side strands of sugar and phosphate and bases A, T, G, and C.

Of particular importance, the two sides of the strands were comple-
ments of each other. When the hydrogen bonds were broken—and they
were relatively weak bonds—a sequence such as ATCCGTA on one side
would have its complement TAGGCAT on the other side. It was easy to see
now how the molecule could replicate. When the hydrogen bonds broke

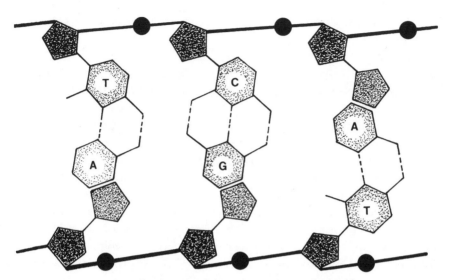

Close-up of DNA bases showing detailed structure.

and the two sides separated, each could form a new and complete molecule. One molecule would become two.

Everything came together and the model was completed on March 7, 1953. News of their success spread quickly and people came to see the model. Wilkins and Franklin arrived from London and were soon convinced that it was an excellent model. Neither had any serious objection to it.

Over the next few days Crick and Watson wrote up a paper for publication in *Nature*. Wilkins and several co-workers, and Franklin and Gosling published papers along with it, confirming the model.

Crick, Watson, and Wilkins were awarded the Nobel Prize for the work in 1962. Franklin, tragically, had died several years earlier, and did not share in the prize, which is awarded only to the living.

The Model

In the Crick–Watson model, DNA is composed of two strands, one wound around the other like a spiral staircase. Between the two rungs of the staircase, holding it together, are the base pairs A–T and G–C (or, equivalently, T–A and C–G). The genetic code is contained in the sequence of bases (e.g., ATCCGAT).

Because of the complementarity, DNA not only can duplicate itself but can do much more. The code contained in its side strands also controls the synthesis of protein, protein that is needed to sustain the cell. It produces this protein via RNA. RNA is different from DNA only in the type of sugar along its spine (ribose instead of deoxyribose) and the replacement of thymine (T) with uracil (U). Uracil and thymine are similar, differing only by a single unit at one of the bonds.

RNA picks up the code from the DNA and produces protein. Proteins are made up from about 20 different amino acids, so that the code along the DNA strand has to specify which amino acids are to be assembled.

There is an important relationship between the three basic molecules DNA, RNA, and protein. They play closely interrelated roles. Information is passed from DNA to RNA; RNA in turn assembles protein, and some of this protein is fed back to the DNA controlling the production of RNA. A simple representation of the scheme is as follows:

It was soon discovered that the code was read in triplets. In other words, a sequence such as ATT specifies a particular amino acid. So each triplet along one of the strands specifies one component of the protein, the total protein being specified by from about 100 to 500 triplets.

After the structure of DNA was discovered, the major problem facing molecular biologists was breaking the code; in other words, finding out which triplet specified which amino acid. A major breakthrough came when Marshall Nirenberg and Johan Matthaei of the National Institutes of Health in Washington, D.C., showed that the triplet UUU specified the amino acid phe. They made up a chain consisting only of U and showed that the resulting protein chain that was formed was phe.phe.phe.... This was the first step, and it was an important one. RNA chains were then made up containing small specified amounts of other bases in addition to U, and over a period of about five years the entire code was broken. Each of the triplets specified a given amino acid. (There was some degeneracy in the code; in other words, more than one triplet specified the same amino acid. UUU and UUC, for example, both code for phe.)

DNA and RNA are the two basic molecules of life, but they are dependent on many things. They need a complex environment in which to reproduce themselves and produce protein—an environment that is supplied by the cell. DNA, for example, needs special proteins or "enzymes" to help it uncoil and unzip so that the code can be picked up by RNA.

Other Life

Now that we know what the basic molecules of life are on Earth, we naturally ask: Do the same molecules form the basis of life elsewhere, on other planets? In other words, would alien life be based on DNA, RNA, and protein as life is on Earth? We have no way of knowing for sure. We do know that the same elements exist throughout the universe (spectroscopy shows no new elements). If we discovered another world, it would have to be made up of the same elements as we see here on Earth, and likely life would have to be centered around the elements nitrogen, oxygen, hydrogen, and carbon. In fact, carbon is critical to life here and it seems that it would have to be the centerpiece of life anywhere in the universe. There is no other element that can form so many bonds so easily and therefore allow the formation of huge chain molecules such as DNA, RNA, and protein.

There is no reason to believe, however, that these molecules would be exactly the same on other worlds. Complex molecules are required for life here, and they would likely be needed elsewhere. It is hard to visualize a form of life that could be based on simple molecules. Furthermore, the mechanisms and relationships we see here would likely have to be the same elsewhere, but the major molecules could have a different form, and likely would.

Another important question is, how did these molecules form here on Earth? Certainly we had the basic components—the atoms. But how did they assemble themselves into DNA, RNA, and protein? And why? We will look into this in the next chapter.

chapter 3

The Development of Life on Earth

Now that we have some familiarity with the basis of life, we are in a better position to look at how life came into being, or more specifically, how the molecules DNA, RNA, and protein were formed. They are complex molecules and couldn't have been created in the Big Bang explosion that created the universe, or even shortly after it. It took time—billions of years—to produce them.

We know these molecules developed on Earth, and it is natural to wonder if they, or something close to them, also developed elsewhere in the universe. If the process that created them here is an inevitable consequence of evolution, it seems reasonable to assume that the same process would occur elsewhere in the universe, assuming the conditions were the same. And if this were true, life would be common elsewhere. In short, alien life *would* exist. We have no idea what form it would take, but we will leave that to later. Our first object is to understand how life formed on Earth. Historically, there have been three theories:

1. Spontaneous generation from nonlife
2. Panspermia
3. Chemical evolution

Let's consider each of them.

Spontaneous Generation

Put forward by Aristotle, spontaneous generation is the hypothesis that was accepted for the first two thousand years or so after the development of civilization. Most of the greats of science such as Newton, Galileo, and Descartes accepted it. People saw meat, stored grain, and vegetables rot, and within a week or so the remains were full of maggots and worms. Life, they believed, was coming from nonlife, and it led most people to believe that this was the way that life first appeared on Earth. It seemed natural and few questioned it.

One who did not accept it was Francesco Redi. Born in Pisa, Italy, in 1626, Redi obtained a medical degree from the University of Pisa in 1647. No one had tested the idea that life came from nonlife; it had just been accepted. In fact, testing or "experimentation" was almost unknown at that time. Galileo had been dead for only a few years when Redi graduated, and his "scientific method" was still largely unknown. Redi was one of the few who was unwilling to accept a hypothesis without testing it. He prepared eight flasks with a variety of meat and vegetable broths; he sealed four of them and left the remaining four open to the air. Within a few days it was evident that the open ones were attracting flies, and as expected they developed maggots. To Redi's surprise, however, the sealed ones did not.

But maybe maggots had not arisen in the sealed ones because air was excluded. Redi wasn't going to take any chances; he carefully laid a fine gauze over the open jars and redid the experiment. No maggots formed in the flasks with the gauze over them, while the others were full of maggots. To Redi it was obvious what was happening. The flies were laying eggs on the decaying matter and the maggots were coming from them. When the flies couldn't get to it, they couldn't lay eggs.

Redi wrote up his results and published them. The idea of spontaneous generation from nonlife was dead. But strangely, it didn't remain dead for long. Within a few years it was back at center stage. It all began with Anton van Leeuwenhoek's invention of the microscope in 1674. With his microscope he could magnify small objects up to 200 times, and with this power a new world was open to him. He examined everything with his new device—rainwater, dirty ditch water, teeth scrapings, hair, and so on. And to his amazement he saw tiny living "animals."

With van Leeuwenhoek's discovery came a resurgence of interest in spontaneous generation. Here was life at a much lower level, and people began to suggest that it arose from nonlife. The English naturalist John

Needham was one of the first to jump on the bandwagon. He sealed mutton gravy in a flask with a cork so that no outside microbes could get in, then heated the flask in ashes. After letting the flask cool he waited for a few days, uncorked it, and examined the contents with a microscope. The gravy was swarming with microbes. The spontaneous generation theory had to be correct. Where else would the microbes come from? He presented his result to the Royal Society of England and within a short time there was considerable excitement about his discovery.

But not everyone was convinced. Lazzara Spallanzani, an Italian biologist, had read about the controversy in Redi's book, and he was not convinced by Needham's experiment. Was it possible that microbes had gotten into Needham's flasks? Had he plugged them tightly enough? Had he heated them enough? Spallanzani decided to do the experiment for himself. He prepared a broth of peas, seeds, and almonds, and began. He considered corks, but decided they weren't good enough so he melted the glass necks of the flasks, drew them out, and sealed them. No microbes could get by the sealed glass. Before he sealed them, however, he applied heat, and he didn't merely put his flasks in hot ashes; he boiled the contents. He boiled the first few for only a few minutes, but the last ones he boiled for almost an hour, then he sealed them.

He left the flasks to cool and waited. Several days later he broke them open and checked the contents with his microscope. He looked first at the ones he had boiled for only a few minutes. Lo and behold, there were microbes present. Not many, but they were definitely there. Then he looked at the flasks that had been boiled for almost an hour. Nothing. There were no microbes. Needham was wrong; he had just been sloppy in allowing microbes to slip into his flasks. He hadn't sealed them well enough, and he hadn't heated them enough.

Although Needham was shocked when he heard the news, he didn't back down. He severely criticized Spallanzani's experiment. "Heat destroys the vegetative force—the force that creates life," he said. "You have heated the broth so long you have destroyed the force."

To Spallanzani it was a stupid argument—he didn't believe in a vegetative force—and he was determined to prove Needham wrong. He looked again at Needham's experiments. Needham had used corks in his flasks; he would do the same. He sealed all his flasks with corks and repeated the experiment, boiling some for only a few minutes and others for almost an hour. When he uncorked them and examined the broth he found that all flasks were loaded with microbes. There were almost as many in the ones that had been boiled extensively as there was in the ones that had only been boiled briefly. The vegetative force—if it existed—obviously wasn't killed by extreme heat.

Still Needham didn't concede. "Spallanzani is changing the elasticity of the air inside his flasks," he said. Spallanzani shook his head and went to work again, and again he proved Needham wrong. But strangely the controversy continued. It came to a climax in 1857 when Felix Pouchet, the director of the Museum of Natural History at Rouen, published a book claiming that he had irrefutable proof that the theory of spontaneous generation was correct.

Finally, the French Academy of Science got into the act; the academy offered a prize to anyone who could resolve the problem once and for all. The challenge was taken up by Louis Pasteur. Born in Dole, France, in 1822, Pasteur was a mediocre student at school, taking little interest in the academic subjects. His main interest, it seemed, was painting, and he was good at it. But his imagination was fired after attending a talk by Jean Baptiste Dumas, the French chemist who showed that there were families of organic compounds. Pasteur cast his aspirations to become a painter aside and decided to become a chemist.

When Pasteur heard of the prize offered by the French Academy he quickly became interested. He had considerable experience working with

Louis Pasteur.

microbes and had often wondered why they appeared so readily in broths that were left in open flasks. Was it possible that there were microbes in the air, possibly on dust? The idea had been around for years, but it had never been checked. Pasteur decided to check it. He put a cotton plug in a glass tube and pumped air through it, then he dissolved the cotton in a sterile broth and examined the remains under a microscope. To his surprise the broth was full of microbes. Microbes were, indeed, airborne—probably on the dust that could be seen drifting in the air. This had to be the way they got into open flasks.

Pasteur began by repeating Spallanzani's experiment. He boiled the broth for varying amounts of time, then sealed them by melting the glass. And he found, as Spallanzani had, that there were no microbes in the flasks that had been boiled extensively. This convinced him that spontaneous generation was incorrect.

"But the microbes need natural air to grow," his critics said. "You are sealing them in."

Pasteur thought about it. How could he get natural air into the flasks, but keep the microbes out? It seemed impossible. Microbes even got in past corks. As he was thinking about the problem, a friend, Antoine Balard, walked into his lab. Balard had discovered bromine a few years earlier. Pasteur told him about his problem. Balard suggested that he should draw the opening out into long tubes, then bend them. The broth in the flasks would be open to the air, but any dust entering the long necks would settle long before it made it into the broth.

Pasteur experimented with various necks, finally settling on one that resembled the neck of a swan. He redid the experiment with the long open necks, and after a few days he broke the necks off and examined the broth. Nothing. There were no microbes. To prove his point further, he broke off the long necks and let the flasks sit. After a few days he looked again at the broth. It was full of microbes. The air in the long necks had obviously acted as a cushion, preventing the dust particles from getting in.

Scientists were finally convinced and within a short time the doctrine of spontaneous generation was dead and gone for good.

Panspermia

With spontaneous generation gone, however, the problem of the origin of life remained. Where did life come from? Perhaps spores drifted through space carrying life, and when they fell on a planet they initiated biological evolution. Many well-known scientists accepted this view, but it wasn't clearly stated until 1908 when the Swedish chemist Svante Arrhenius gave it a name—panspermia. He proposed that spores (bacteria wrapped in protein) escaped from planets that contained life and drifted through space, propelled by starlight, and as they approached a star they were attracted to it by gravity. Some of the spores then collided with planets orbiting the star. Earth was one of those planets.

If this is, indeed, true, it still leaves us with the problem of the origin of life—where the spores came from. One possibility is that life was generated when the universe came into being; in other words, it's a property of the universe, like matter and energy. As we saw earlier, this is extremely unlikely. Life is too complicated. The basic molecules are complex, and it takes time—eons—to assemble them from their components. It's not something that happens overnight.

Aside from this, though, there's a serious problem with panspermia. We now know much more about distances in space and the effects of radiation and temperature on spores than we did when Arrhenius put the idea forward. The universe is full of radiation—X rays, gamma rays, and UV light—which would be lethal to spores traveling through it. They would need considerable protection from it, if they were to survive. Furthermore, the distances they would have to travel are so great, and the corresponding times the spores would have to spend in space is so long, it is extremely unlikely they could survive. But even more detrimental to the theory is that if it were true we should have found spores all through the soil on the moon and Mars, and we didn't. The rock and soil samples brought back from the moon contained nothing of biological interest, and the *Viking* craft on Mars showed no evidence of spores in the Martian soil (although, as we will see, evidence of life was found recently in a meteorite from Mars).

Despite the evidence against it, panspermia is still not dead. Fred Hoyle of Cambridge and a collaborator, Chandra Wickramasinghe, resurrected the theory in 1979 by pointing to optical evidence (not accepted by most astronomers) that interstellar dust grains are made up of bacteria and algae. There is no doubt that there are many chemicals in space—some important to life, and some associated with the basis of life—but there is no evidence that there are actual *life forms*.

Francis Crick, one of the discoverers of the structure of DNA, believes that the problems in trying to show that intelligent life arose naturally on the primitive Earth are so great that life must have come from space, and he has put forward what he calls directed panspermia. According to his theory, spores were purposely directed at targets in space by an advanced civilization. Crick has teamed up with Leslie Orgel, a well-known experimenter on the origin of life, in putting forward this idea. They argue that space should be full of advanced civilizations, and that some of them deliberately seeded planets around the stars they visited. Earth was presumably among them.

There is no evidence for this; on the other hand, there is no evidence against it. It would, in fact, be very difficult to prove one way or another. Fortunately, there is an alternative to panspermia, and it is the one accepted by most scientists.

Chemical Evolution

Chemical evolution is a distant cousin to spontaneous generation. Life is again evolving from nonlife, but it is not doing it instantaneously in the form of microbes. Life, according to this theory, came about as a slow evolutionary process. It did, indeed, arise from nonliving matter, namely, atoms and molecules, but the process of building up the necessary macromolecules took billions of years.

The first theory of this type was proposed by the Russian biochemist A. I. Oparin. Born in Moscow in 1894, Oparin graduated from the University of Moscow in 1917. His interest in the origin of life began in the early 1920s; he published his most famous book, *The Origin of Life on Earth*, in 1924. It was translated into English in 1938. Since Pasteur had squashed spontaneous generation, few people had speculated on life. It was a delicate subject, and anyone working in the area had to tread carefully. Oparin surveyed the field thoroughly before publishing his ideas. He began by postulating that the primitive atmosphere of the Earth was different from its present one, consisting mostly of methane, ammonia, and water vapor, with some hydrogen. No oxygen would be present because photosynthesis had not yet begun (it requires plant life).

He based this conclusion on several lines of evidence. Hydrogen, for example, was the most common element in the universe, and methane and ammonia were prominent in the atmospheres of Jupiter and Saturn. He proposed that the basic units of life would arise naturally from this atmosphere as energy was applied to it, and they would interact, pro-

A. I. Oparin.

ducing complex molecules. Eventually these molecules would give rise to life.

A similar theory was put forward by the British biologist J. B. S. Haldane in 1929. Like Oparin, he envisioned a primitive atmosphere consisting mainly of methane and ammonia, and he assumed life molecules would form naturally as energy from the sun and other sources was applied to it. He went a little further than Oparin, however, in that he suggested that these molecules would accumulate in the oceans until they became like a "hot dilute soup." The buildup to complex life forms would occur here, and eventually they would migrate to the land.

Strangely, as late as 1952 these ideas were still generally unknown to most scientists. In the United States Harold Urey of the University of Chicago came to the same conclusion independently. Born in 1893, Urey majored in zoology at the University of Montana, and he obtained his Ph.D. from the University of California in 1923. He is best known for his discovery of heavy hydrogen for which he received the Nobel Prize in 1934. Urey's major interest was not the origin of life but the origin of the solar system, the planets in particular. In working out the details of planetary formation he came to the conclusion that the Earth had to have a reducing atmosphere consisting of ammonia, methane, water vapor, and hydrogen. It is referred to as "reducing" because of the presence of free

Harold Urey.

hydrogen and other molecules that readily combine. According to Urey's theory, this reducing atmosphere gradually changed to an oxidizing one as life began to flourish on Earth. With hydrogen being a light element it would escape to space easily, and it was during this period of escape when the atmosphere was in a mixed oxidating–reducing state that most of the formation of complex life molecules occurred.

Before we get into the details of how this happened, let's consider the geological evidence. We know that the Earth is approximately 4.5 billion years old. Several different ways of dating it—radioactive decay, salinity of the oceans, and so on—are known, and they are all in general agreement. Furthermore, we know that a primitive form of life existed on Earth 3.5 billion years ago: fossilized imprints of bacteria have been found in rocks that are 3.5 billion years old. This seems to indicate that life formed during the first billion years, but for the first half billion or so our planet was extremely inhospitable to life; it was battered by comets and asteroids, and volcanic eruptions and violent storms were common on its surface. That leaves only about 500 million years for life to form. This is only approximate, however, so to be on the safe side we'll say from 300 to 700 billion years. This may seem like a long time, but it's actually very short. During this time the basic molecules of life had to form and assemble themselves into DNA, RNA, and protein. Furthermore, cells had to be produced, and bacteria such as that found in fossilized form had to come into being. So the window for the origin of life is narrow, and this is still a problem to many scientists.

Since Urey's time another problem has also developed. One of the major reasons for the hydrogen-rich reducing atmosphere, according to Urey, was that the Earth originally had its iron uniformly distributed throughout it. This meant that there was considerable iron near the surface, and according to Urey, this iron reacted with water to release hydrogen. Later, because of its high density, the iron sunk to the core of the Earth where it is today.

Many geologists do not agree. They believe the iron condensed to the center while the Earth was still forming from the solar nebulae. This would mean that the Earth's primitive atmosphere was considerably less reducing than Urey supposed. Methane and ammonia would be present in limited amounts, if at all. Carbon monoxide and carbon dioxide would be more common. As we will see, however, this does not pose a serious problem.

Others may have had the idea before Urey, but they didn't do anything about it; they didn't test it. Urey did. He and graduate student Stanley Miller performed an experiment that is now considered to be a classic. As the story goes, Miller went to a lecture on the origin of life by Urey at the University of Chicago. Urey mentioned that many people had tried to create life from nonlife, but had failed. The reason, he said, was that they were using the wrong atmosphere.

Miller went up to Urey after the lecture and asked him if he could do this experiment as a thesis project. Urey agreed, and Miller built the apparatus. He used a simulated atmosphere of methane, ammonia, water vapor, and hydrogen. Steam was used to circulate and trap by-products of the chemical reactions. Energy was supplied to the system by electrodes, similar to lightning on the early Earth. The sun was probably the major source of energy, particularly the ultraviolet light from it, but lightning would have also been important.

At the end of a week Miller analyzed the chemical products and was amazed to find several biologically important molecules. In particular, four amino acids were present: glycine, alanine, aspartic acid, and glu-

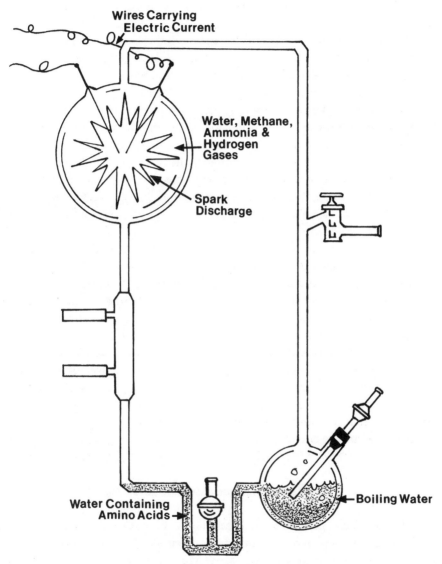

Miller's apparatus used in simulating the primitive atmosphere.

tamic acid. Of interest, several amino acids that are not found in protein were also produced. In addition to amino acids, however, other molecules of biological interest were generated, such as hydrogen cyanide and formaldehyde.

Miller's experiment was the first and it led the way. Over the next few years Miller continued to perfect his techniques and his apparatus, and as you might expect, several other groups soon jumped on the bandwagon. In these later experiments slightly different primitive atmospheres were used, along with different energy sources. A mixture of carbon monoxide, carbon dioxide, nitrogen, and water vapor along with a small amount of hydrogen was tried, and like Miller's atmosphere, it produced amino acids and other biological molecules. As long as the atmosphere was reducing, life molecules were produced. Oxidizing atmospheres, on the other hand, produced nothing of biological interest. Other energy sources such as ultraviolet light, shock waves, and particles from accelerators were also tried, and all were effective.

So far, though, we've focused on the components of protein, namely, amino acids. What about the components of DNA and RNA, in other words, sugars, phosphate, and the bases A, T, G, C, and U? Some but not all of these compounds have been obtained in Miller-like experiments. In these experiments biological molecules such as hydrogen cyanide and formaldehyde—compounds produced in the Miller–Urey experiment—were used as starting sources. Juan Oro and Leslie Orgel have obtained all four bases of RNA in experiments of this type. The only base not obtained so far is thymine (T). Experiments have also shown that the sugars ribose and deoxyribose can be obtained in a similar manner. Phosphate is not a problem in that it would have been naturally abundant on the early Earth in rocks and in the oceans.

Other Routes to Life

Although it is generally assumed that most if not all of the basic units of the large molecules of life came from reactions in the primitive atmosphere, it is possible that some of them came to Earth by another route. They might have come to Earth via comets and asteroids. Collisions played a central role in the buildup of the planets. They were no doubt much more common in the early stages of the solar system's development than they are today. Collisions within the solar system are relatively rare now, but we did see a dramatic series of collisions with the planet Jupiter in 1994 when pieces of Comet Shoemaker–Levy 9 hit it.

Anyone doubting that collisions were common in the past need only look at the surface of the moon. It is pocked with craters and scars of all sizes. We see a small number of craters left by collisions on Earth, but they are rare compared to the moon because erosion erases all traces of a crater on Earth in a relatively short time, about 25,000 years. Craters that occurred 2 and 3 billion years ago are gone here, but they are still visible on the moon.

Collisions are important because asteroids and comets are known to contain biologically important molecules. Carbonaceous chondrite meteorites, a class that makes up about 5 percent of all meteorites, contain about 2 percent organic matter along with considerable water. They are objects that were formed with the solar system 4.5 billion years ago. Amino acids have been found in several meteorites. The Murchison meteorite, which fell in Australia in 1969, has been shown to contain eight different amino acids, along with the bases adenine, guanine, and uracil. Comets have also been shown to contain biological compounds.

On the basis of this it is possible to infer that the Earth was seeded to some degree from space. This is not the panspermia we discussed earlier, which assumes that large, intact biological molecules such as DNA, RNA, and protein came from space. In this case we are only assuming that the precursors of these molecules (i.e., amino acids and the bases ATGC) came from space; furthermore, we are talking about the space within our solar system.

We will likely never know the extent of this seeding, and how it compares to what the Earth itself produced, but it does seem that asteroids and comets may have made a contribution. And indeed if they did, we wonder how these molecules were formed in space where conditions are much more hostile to life than they were on the early Earth. It may be that they form more easily than we think.

Another clue to the origin of life is the large number of molecules that have been found in space by radio astronomers. Many different organic molecules have been identified in large clouds of gas in interstellar space. These clouds are similar to the clouds that made up the solar nebula from which our solar system was formed. Although nothing as complicated as an amino acid has been found yet, many biologically important molecules such as carbon monoxide, cyanogen, ethyl alcohol, and methylamine have been identified.

Putting the Pieces Together

So far we've only shown that the components of DNA, RNA, and protein would have formed naturally on primitive Earth. We still have to look into how these components came together to give us these molecules. We will see that this is still not fully resolved.

Let's consider protein first. It might seem that it would be a simple matter to produce protein from a mixture of amino acids. Start with a given amino acid and add another amino acid to it, then another, and so on. All that is needed is a small amount of energy and the elimination of a water molecule. At one end of an amino acid is an atom of oxygen, and at the other are atoms of oxygen and hydrogen, so when they come together water is formed. For the two amino acids to join, the water molecule must drop out of the structure. What is left is a peptide bond.

It sounds easy, but there's a problem: the reaction is reversible. If water is available, the bond is easily broken, and the two amino acids separate. So it is important that once the bond is formed it is removed from the water, or at least protected from a break.

With DNA and RNA things are more complicated, but again to form a chain of nucleotides (the basic component of DNA or RNA) you need energy and the release of a water molecule. Within the cell this is easily accomplished. Protein, acting as an enzyme, supplies the energy and provides the environment so that water is released. Outside the cell it is difficult to say how this would proceed. The first thing we would need is a concentrated solution of the components. But even with a concentrated solution, proper bonds would not form without some guidance. There are roughly 100 different possible bondings that could occur between the various components, and they would likely occur randomly. The result would not be DNA or RNA.

Something else that affects all three of the basic molecules is ultraviolet light; it is needed as an energy source to form the components, but it also breaks them up. Therefore, once the basic molecules are formed, they need protection from it. Scientists believe that they got protection from water. The torrential downpours that were common at the time would have washed many of these molecules into the oceans where they would be safe from ultraviolet rays. But this gives us a catch-22. The components would be safe from ultraviolet rays, but they couldn't come together to form chains. As we just saw, this requires the release of water, and it's virtually impossible for a molecule to release water when it's surrounded by water.

One way around this dilemma is to assume that ponds and lagoons at the edges of the oceans played a role. Tides would also have been helpful. J. D. Bernal of the University of London was the first to point out their importance. Clay at the bottom of the ponds may have helped in assembling the molecules. Organic material adsorbs to clay. The evaporation of ponds along with freezing would help concentrate the components, and as the first life forms emerged (particularly blue-green algae), photosynthesis would begin, and an ozone layer would soon form in the atmosphere, giving life forms that migrated to the land protection from the ultraviolet radiation.

Once the three basic molecules of life were formed, however, something was still needed: a cell to house them. In particular, they needed the conditions and environment of the cell to begin interacting. Scientists have considered two possible routes to the cell. Oparin showed that under the right conditions a mixture of organic material and water will separate, with the organic molecules forming small spheres called coacervates, which have some of the properties of cells. They can, for example, grow and split. Furthermore, they can take in substances from the aqueous solution around them.

Sidney Fox of the University of Miami has shown that chains of amino acids, which he refers to as proteinoids, will spontaneously form spheres called microspheres, which also have many of the properties of a cell. In particular, they can absorb substances from the solution around them. Microspheres and coacervates are a long way from the cells we are familiar with, but they may have been the first step.

The Chicken or the Egg?

Almost everyone has heard the age-old question: Which came first, the chicken or the egg? And we, of course, have the same problem in relation to the origin of life. Which came first: DNA, RNA, or protein? Inside the cell there is a complex interaction between these three molecules, and each depends on the others to fulfill its function. Let's begin with a quick review of this interaction. As we saw earlier the two strands of DNA separate, and from these strands, two new DNA molecules can be constructed. But DNA also directs the construction of protein. For this, an RNA molecule is needed, and it is constructed along the same lines as DNA replication. RNA is complementary to the DNA from which it is formed, with U along the chain wherever T appears in DNA. It is referred to as *messenger RNA*, and the sequence of nucleotides along the RNA specifies a sequence of amino acids that will make up a protein. Three successive nucleotides specify one amino acid.

In addition to messenger RNA there are two other types. *Transfer RNA* carries the amino acids to a site called the ribosome. At the ribosome a third type of RNA called *ribosomal RNA* helps form the peptide bonds that join the amino acids together. All in all, it is a complicated process, and it seems unlikely it could have sprung up accidentally.

This brings us back to the question: Which of the three molecules came first, and how did it produce the others? Most scientists now believe that RNA came first. This is the point of view put forward by Manfred Eigen in the 1970s. RNA is a particularly viable molecule. We now know that it can replicate without the help of DNA, and furthermore, it can act as an enzyme, assisting some of the chemical reactions that take place in the cell. Because of this, many scientists now believe that there was an early stage on Earth in which only RNA existed (no DNA or protein was present). All the functions of the rudimentary life that existed at that time were presumably carried out by RNA. These early RNA molecules were probably short, containing no more than about 60 to 80 nucleotides, compared to thousands in most of the RNA molecules of today.

Opinions vary as to what happened next, but most scientists believe that the chemical machinery for protein synthesis was developed by the RNA, then it somehow encoded the genetic message onto another long molecule that eventually became DNA. These three molecules were later incorporated into cells.

Another point of view was suggested by Oparin in his book *The Origin of Life on Earth* in 1924. Oparin, who experimented extensively with coacervates, believed that they were the precursors of modern cells. He proposed that they formed first; then came the protein in a process similar to that in the Miller–Urey experiment. The protein migrated to the coacervates, and eventually RNA and then DNA were produced within the primitive cells. This picture was accepted for several decades after Oparin suggested it, but as more and more was learned about the three basic molecules, it seemed that RNA and DNA were more basic than protein and the cell.

A third theory by A. G. Cairns-Smith of Glasgow is similar to Eigen's, but with some basic differences. His theory goes back to the assembly of DNA and RNA. How did they assemble themselves from their building blocks? Cairns-Smith believed they were assisted by clay. Since clay can form long crystal arrays, it seems as if these chains may have initially been

a template for RNA, helping it assemble itself. And, as faults and disloca-
tions occurred in the clay crystal, RNA would mutate and change, until
finally it began to function independently of the clay. According to this
theory, protein came after RNA and DNA, and cells came last.

All of these theories are viable, but at the present time most scientists
believe Eigen's theory is the correct one.

The Rise to Intelligence

There are many unresolved problems in the story I have just told. Still,
it is the best understood part of the rise from chemicals to intelligence.
Much of what we know beyond the first cells is based on speculation.
Fortunately, we have fossil remains to guide us.

Viruses are the simplest life forms, but they need cells to function, so it
seems unlikely that they came first. Prokaryotes, or cells without nuclei,
likely came first. They contain RNA and DNA and have a relatively
complex reproductive mechanism. Basically, they are bacteria. The earliest
fossil remains—again, about 3.5 billion years old—are of this type. A
particular type—blue-green algae—produced the oxygen in our atmosphere via photosynthesis, and with it came the ozone layer that gave
protection from the ultraviolet radiation from the sun. The transition from
a reducing atmosphere to an oxidizing one occurred between 2 billion and
2.5 billion years ago.

Once there was free oxygen in the atmosphere a more complex cell
appeared—a cell with a nucleus, called an eukaryote. It appeared between
1 and 2 billion years ago. These cells contain up to 1000 times as much DNA
as prokaryotes.

From the first eukaryotes the evolution to multicellular organisms
began. The first eukaryotes appeared about 1.6 billion years ago, then
about 600 million years ago there was a sharp increase in the number and
variety of organisms. This was the beginning of what is called the Cam-
brian era. A particularly abundant fossil of this era was the trilobite, a
distant relative of the scorpion. From here the rise to dinosaurs was quick;
they existed about 200 million years ago. Beyond dinosaurs there are a lot
of uncertainties, but we believe the first humanoids appeared about 3
million years ago.

With our understanding of how life formed on Earth we are now in a
position to consider the possibility of life beyond Earth, and in particular, if
there is life, what form it might take.

chapter 4

Alien Life

What would aliens look like? Science fiction writers have been trying for years to imagine them, but in reality no one knows for sure. You might think that anything is possible, and to some degree that is true. But there are limits, particularly in relation to intelligent life.

We can get some idea of the possibilities by considering the diversity of life on Earth. It ranges from tiny microbes that are barely visible even when magnified 100,000 times, to giant redwoods and blue whales. Though redwoods and microbes are both alive, they are quite different forms of life and bear little resemblance to one another. Furthermore, there are millions of species between them that are all quite different.

The number of species on Earth is, indeed, immense, but strangely only one—humans—has developed a high intelligence. Would we expect the same on another planet—in other words, millions of species and only one with high intelligence? It's hard to say. It is reasonable to assume that some of the species on a planet with life would develop intelligence, but we could never be certain how many. Furthermore, even if they were intelligent, they might not have a technology. Dolphins have a larger brain than humans, and there's considerable evidence that they are fairly intelligent, but they haven't developed a technology.

Despite its diversity, all life on Earth does have something in common. It is dependent on the molecules DNA, RNA, and protein. The genetic code of the tiny bacteria is the same as that of the giant redwood. Furthermore, all life on Earth is based on carbon, and the solvent, water, that sustains life is the same everywhere. Would this be the case on another planet?

Carbon-Based Life

Everyone is familiar with carbon in one form or another. Coal is mostly carbon, as are soot, charcoal, and graphite, so you merely have to look at the graphite in your pencil to see a good example of it. And with this many examples around, it's obviously a common substance. This is important, as life would have to be based on an element that is readily available. But we need more than this. What else does carbon have that makes it so ideal in relation to life? We need only look at its structure to see. It has four bonds, the maximum number any element can have, and because of this it can form many compounds.

Let's compare carbon to some of the other elements on Earth, elements such as hydrogen, oxygen, and nitrogen, all of which are abundant and important in relation to life but are not the basis of life. We can represent the four bonds of carbon as follows:

$$-\overset{\displaystyle |}{\underset{\displaystyle |}{C}}-$$

Other elements can attach themselves at each of the bonds. Hydrogen, for example, has one bond and if a hydrogen attaches itself at each of the bonds we get methane.

$$H-\overset{\displaystyle \overset{H}{|}}{\underset{\displaystyle \underset{H}{|}}{C}}-H$$

What is particularly important, though, is that carbon atoms bind easily to one another, giving chains of atoms. Furthermore, if hydrogen atoms attach themselves to each of the bonds we get compounds such as

$$H-\overset{\displaystyle \overset{H}{|}}{\underset{\displaystyle \underset{H}{|}}{C}}-\overset{\displaystyle \overset{H}{|}}{\underset{\displaystyle \underset{H}{|}}{C}}-\overset{\displaystyle \overset{H}{|}}{\underset{\displaystyle \underset{H}{|}}{C}}-\overset{\displaystyle \overset{H}{|}}{\underset{\displaystyle \underset{H}{|}}{C}}-H \;\cdot$$

In this case we have a chain of four carbon atoms, but longer chains are also possible. In fact, compounds containing dozens of carbon atoms are common; the branch of chemistry called organic chemistry is based on them. Furthermore, other elements such as oxygen and nitrogen can also attach themselves at the bonding sites.

If we look at other common elements such as oxygen, nitrogen, phosphorus, and sulfur we see that long chains of this type do not form. Nitrogen, for example, has three bonds, and if hydrogen atoms attach themselves to these bonds we get ammonia.

$$H-N-H$$
$$|$$
$$H$$

But this is as far as we can go. Nitrogen does not form long chains, and as we saw earlier, life requires complexity. Long molecules such as DNA, RNA, and protein are needed.

Carbon is versatile because it can form the maximum number of bonds—four. If any other element were to act as a basis of life it would have to have the same ability. But what about an element out in space that we are not familiar with? We needn't worry about this. Spectroscopic analysis has shown that the elements throughout the universe are the same. There are no elements that are unknown to us, so life on some distant planet couldn't be based on an unfamiliar element.

If we look along the periodic table we find only one other element that can form four bonds—silicon. Could life be based on silicon? Let's look into this.

An Alternative: Silicon-Based Life

First we must ask: Is silicon abundant enough on Earth to support life? It is; in fact it's 135 times more abundant than carbon. If we look out into the universe, however, we find that silicon is not particularly abundant; considerably less than 1 percent of the universe is made up of it. Nevertheless, because it is common on Earth, we could assume that it would be common on planets beyond the solar system.

When we look closely at silicon, however, we see a problem: the bond between two silicon atoms has only one-half the strength of the bond between two carbon atoms. Furthermore, the bond strength of silicon and hydrogen, and silicon and oxygen is much higher than between two silicon atoms. This means that chains of silicon atoms are relatively unstable, and would easily break. In addition, silicon's affinity for oxygen is a problem. In water, silicon oxidizes to form silicon dioxide, which is quartz. This means that water cannot be the solvent in a silicon-life world.

There may be ways around some of these difficulties, but they lead to situations that are very unlikely. Furthermore, if silicon were the basis of life elsewhere we should find some evidence of it in the sky, and we don't. We find hydrocarbons of all kinds in comets, meteorites, in the atmospheres of planets in our solar system, and in interstellar clouds in space, but we find little or no evidence of silicon compounds. On the basis of this we can say with some confidence that life elsewhere in the universe is likely based on carbon.

The Problem of a Solvent

Life elsewhere in the universe would need a solvent to deliver nutrients to the basic molecules, carry off wastes, and regulate temperature. Water performs these functions on Earth, and we know it is an excellent solvent. But would it have to be the solvent on another world? Let's begin by considering what is required of a solvent. First, it would have to remain liquid over a large range of temperatures and would have to have the ability to dissolve certain chemical compounds. Water remains liquid over a hundred degrees (centigrade), and the temperature range over which it is liquid is ideal. Most chemical reactions occur in this range; furthermore, it is not so high that collisions would break up and destroy molecules or so low that life could not be sustained. Furthermore, water has a high heat of vaporization, which means that it takes a lot of energy to vaporize it. This puts a check on sudden changes in temperature, minimizing and sheltering life in the water.

Another important property of water is that it expands when it freezes instead of contracting. Because of this, ice forms on the tops of lakes and ponds, and icebergs float. If water froze and fell to the bottom of lakes all life in them would soon cease.

If we compare water with other possible solvents, such as alcohol or ammonia, we see that it is vastly superior. Ammonia, for example, is liquid only at well below zero ($-78°C$ to $-33°C$) and the size of its range (45 degrees) is only half that of water. Alcohol has a better range ($-94°C$ to $+65°C$) but has many other properties that make it much less desirable than water; water, for example, has twice the ability of alcohol for carrying molecules in solution, and it has a much higher heat of vaporization.

Another plus for water is that it helps protect life from radiation. Most stars radiate in the ultraviolet, and life forming on a planet around it would eventually need protection from this radiation. When water evaporates, some of the molecules are dissociated by ultraviolet light to form ozone. This ozone collects in the atmosphere, forming a shield from the ultraviolet light. Neither ammonia or alcohol gives protection of this type.

Water is obviously a particularly desirable solvent, and again, it is likely that life on a distant planet would be dependent on it.

The Form of Aliens

Even though it's reasonable to assume that life throughout the universe is based on carbon and uses water as a solvent, we are still at a loss as to what this life would look like. About all we can do is guess, but speculation is always fun, so let's give it a try. We'll begin by assuming that the planet we are considering is similar to Earth in that it has large land masses, moderate temperatures, and oceans. It's unlikely the life would be based on DNA, but it would have to be based on a similar molecule, and there would likely be other large molecules such as RNA and protein to assist it. Furthermore, mutations would likely occur, and natural selection— with only the fittest surviving—would be important, as it has been here on Earth.

We are, of course, mainly interested in intelligent life, so we'll restrict our discussion to it. On Earth, neglecting dolphins, we are the only intelligent species, so it's logical to begin with us. What can we say about ourselves? First, we are relatively small compared to many species on Earth, particularly those in the past such as dinosaurs. We are mobile, motivated, and we try to accomplish things—mostly via technology. This would tend to rule out plants, low forms of life, and extremely large species as intelligent beings.

There is, of course, no reason to assume that intelligent aliens could not be considerably smaller, or larger, than humans, as size and muscular structure depend to a large degree on the gravity of the planet. We have two arms and two legs. Would this necessarily be true of aliens? Many of the creatures on Earth are four-footed; humans, or at least a primitive ancestor of humans, at one time likely used all his limbs as feet. Gradually, though, he stood upright and used two of them as arms. Aliens may or may not stand upright, and it is possible that they would have four legs and two arms. Fingers are also important to us. Aliens would likely have fingers, or something resembling them—perhaps tentacles.

Another feature we have is a head perched on the top of our body. This allows us to move it around so that we can easily observe what is around us. Furthermore, our brain is encased in a strong, bony shield to protect it. The brain of an alien would also likely be encased in a shield to protect it, but it wouldn't necessarily be located on the top of the body. It could be anywhere. It is also possible, of course, that an alien could have more than one brain, or one that is not centrally located—it could be spread through the alien's body.

Since we are dealing with a hypothetical planet with oceans we should also consider the possibility of intelligent ocean life. Life on Earth began in the oceans, and perhaps by lucky chance the species that emigrated to land developed the higher intelligence. The ocean depths provide a calm environment; temperatures are relatively stable and there is not the turbulence (storms) that occur on the land.

On Earth the dolphin has risen to a relatively high intelligence. Its brain is larger than a human's, and convoluted to the same degree, and we know that dolphins can communicate with one another. In fact, even though we do not understand their language, we know they are amazingly intelligent in that they can solve mazes and puzzles. Because they have fins (which they need for swimming) rather than fingers, an ocean-based intelligence would have limited tool-making capacity and could not develop a technology.

Another possibility is an intelligent form of life that is able to fly. If we look at the airborne creatures here on Earth, though, we see a stumbling block. All birds have relatively small brains, compared to their body size. Nature has economized because flying is difficult and any excess weight would be a problem. An excessively large head on a flying alien is not out of the question, but it presents problems and is therefore unlikely.

One place where we see a lot of speculation is science fiction: aliens of all types and sizes are seen on the pages of science fiction books, in movies, and on television. Do they give us any insight into what aliens might be like? Let's take a look.

The term *science fiction* was coined in 1851, and many consider 1857 to be the year in which this form of fiction was born. H. G. Wells gave us some of the first aliens in 1898 in his *War of the Worlds*. Little was known about Mars at the time, so almost anything went. Wells's aliens were telepathic, wore no clothing, and had tentacles. According to his description, "They had huge round bodies—or, rather, heads—about four feet in diameter, each body having in front of it a face. The face had no nostrils, but it had a pair of very large, dark-colored eyes, and just beneath this a kind of fleshy beak." The tentacles, 16 in number, emanated from near the mouth. All of this, we now know, is unrealistic because we know much more about Mars than we did then, but it was imaginative for the time.

Edgar Rice Burroughs, the creator of Tarzan, also wrote a series of books about Mars early in the century. He was no doubt inspired by Percival Lowell's prediction that Mars was inhabited. Burroughs's first book, called *Princess of Mars*, had a variety of strange-looking and hostile Martians in it, but amazingly, the scantily clad Martian princess looked like the heroine of a B movie on Earth.

Not all aliens were menacing and strange, however. I remember the impact the movie *The Day the Earth Stood Still* had on me in the early 1950s. The alien, Klaatu, was as human looking as any human. The only strange thing about him was that he had a robot companion called Gort. Klaatu came to Earth to warn us about our destructive ways. In this case it was the Earthlings who were menacing; they shot him.

Also from the 1950s was Hollywood's production of Wells's *War of the Worlds*. We only got a glimpse of the aliens themselves—they were similar to those described in the book—and they were menacing. The special effects were impressive, with the invincible Martians almost overcoming Earth's meager defenses. Also from this era came *The Invasion of the Body Snatchers*, with its touch of panspermia. In this case the aliens came to Earth as spores. Once on Earth they germinated into giant seedpods, made duplicates of the bodies of humans, and finally took them over. It was an ingenious plot that no doubt scared a lot of people.

Menacing aliens were also present in the 1954 production of *This Island Earth*. In this case they had large, bulbous heads and clawlike appendages, but in other respects they were like humans. A variety of aliens were presented in *Invaders from Mars* (1953), from the crablike leader to the potato-men and pseudodinosaurs. They implanted devices in the necks of Earthlings in their effort to take the Earth over, but of course, they didn't succeed.

The launch of the Russian satellite *Sputnik* in 1957 gave science fiction a boost. Television was now in most homes, and in 1966 came the first episodes of a series that would have a tremendous impact on science fiction for years. The series was "Star Trek," with the spaceship *Enterprise*. The leader was the very human Captain Kirk, and the science officer, the

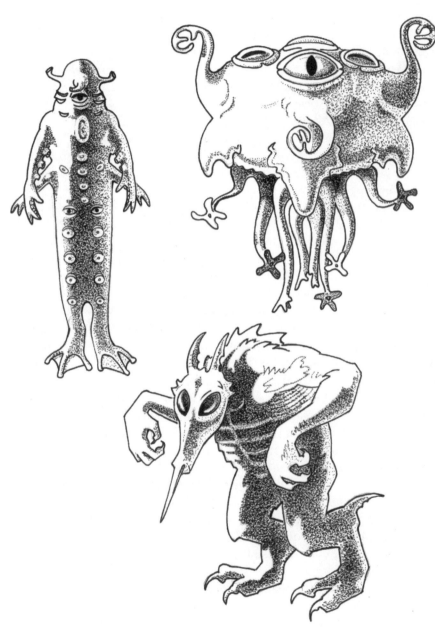

Aliens from the pulp novels of the 1940s and 1950s.

half-human, half-alien Mr. Spock. The show ran for only two seasons, and it was not a big success, but it developed a loyal following and under pressure from its fans it was brought back for another season after it was cancelled. Its real success came in reruns, and of course in recent years there have been sequels to the series with new casts. If you want to see aliens you merely have to watch a few episodes. Over the years at least 200 different life forms have crossed the screen, including the menacing Borg, the Klingons, and the Romulans. Although many of them have strange features, they are all humanlike, and all (as you might expect on television) speak English.

In 1968 came Stanley Kubrick's amazingly successful *2001: A Space Odyssey*, based on Arthur C. Clarke's novel *The Sentinel*. It was a puzzling and strange movie to many, but curiosity and word of mouth made it one of the most successful science fiction movies. Aliens were presented only indirectly; none appeared on the screen. Nevertheless, it was a story of alien contact.

The first landing on the moon came a year later in 1969 and it inspired several excellent science fiction movies. One of the most successful was George Lucas's *Star Wars*. Aliens of all types were presented, ranging from the three-foot-high Jawas who traveled the wastes of the planet Tatoone collecting scrap, to the Tusken Raiders they sold the scrap to. The Jawas were rodentlike beings, clothed in rough, dirty cloaks, that communicated in low gutteral hisses. The Tusken Raiders were almost as repulsive (to us); they were large, strong, and lived in Tatoone's desolate regions. The large range of aliens on the planet was seen strikingly in the cantina scene; there was an amazing and unforgettable array of them.

The same year—1977—that gave us *Star Wars* gave us Steven Spielberg's *Close Encounters of the Third Kind*. It was a highly successful film in which Earthlings had their first encounter with aliens. According to Spielberg's description, the aliens "seemed about three feet high, humanoid in that they had arms and legs and a kind of bulbous head.... Their arms and legs were incredibly flexible in a way we humans could not imitate." Unlike many of the earlier aliens, they were not menacing.

Spielberg continued on with his even more successful *ET*. Again we had a nonmenacing alien, quite different from humans. It was a heartwarming movie that was a tremendous hit with children.

But that was far from the end of menacing aliens. The year 1979 brought us one of the scariest science fiction movies, *Alien*. According to the description given in the book, the alien "looked like the hand of a skeleton with many fingers curled into a palm. Something like a short tube protruded from the palm and something like a tail was coiled beneath the base of the hand. There, on the back, was a dim, convex shape like a glazed-over eye." The creature was enough to scare anyone.

Finally, more recently we had the movie *Independence Day*. The aliens had both fingers and tentacles and communicated via telepathy. They had strange-looking heads, but the heads were on top of bodies that were similar to ours.

What can we say about the aliens in science fiction? Are they plausible? It's obvious that a lot of imagination has gone into them, and we can say that more modern aliens—those in recent science fiction—are more plausible. Almost anything you can think of has been used in science fiction at one time or another. Overall, that gives us a lot of aliens to ponder, but even after looking them over we're no closer to what a real alien might look like. We have some general guidelines, but I don't think we'll know for sure until we actually meet one.

Nonchemical Life

So far we have talked about life based on carbon, but it is possible to have life that is not based on chemical interactions. Three possibilities are computer intelligence, gaseous life such as might exist in a large gas cloud in space, and life on a very large scale—the scale of galaxies—where stars are the basic units. Let's begin with computers.

Computer intelligence would, of course, have to begin with biological life. Somebody has to build the computers. But if we want to stretch our imaginations, it is possible that computers could break off and form their own civilizations. Over the past few years computers have become extremely sophisticated. How much further can we go? Will computers, for example, ever be able to think for themselves? They are in many ways much more efficient than humans in that they can do things humans can't do. They can, for example, crunch millions of numbers in microseconds. As hard as it is to believe, though, there are limits to what computers can do. The problem of listing all possible routes for visiting 100 cities in the United States is one that is, surprisingly, beyond computers, because it is a problem that requires 100! ($100 \times 99 \times 98 \times 97 \times \ldots$) or more computations. Scientists have shown recently that there are things that the human brain can do that computers will never be able to do, things such as devising a new theory. A computer could never come up with the theory of general relativity, for example.

This doesn't mean that in the very distant future computers couldn't become as intelligent as humans. Fortunately, the scenario of computers getting out of control and taking over the world with humans as their slaves is unlikely. A more likely scenario is one in which humans begin to link with computers via virtual reality and then like it so much that society is changed in that direction.

But if humans are likely to remain masters over computers, how could a computer civilization form? One possibility is that a disaster such as a nearby exploding star extinguishes all life on a planet, leaving computers on their own. This, of course, brings us back to the question of whether computers could become intelligent enough to develop their own society. We can build computers with a large amount of memory, and if this were all there was to it, computers could eventually duplicate the human brain. But we know that things are not this simple. The main problem at the present time is that we still don't understand our own thought processes, so it is difficult to predict whether computers will eventually be able to think. Let's go out on a limb, though, and assume that it is possible. What would the resulting society be like? Computers obviously have a lot of advantages over humans. If they become "sick," it's merely a matter of replacing a circuit element. Furthermore, humans age and die, and with their death goes all the skills and intelligence they have developed over their lifetime. This is not the case for computers. In theory, computers can "live" forever, as long as they are maintained; in contrast, a new baby starts from scratch. Babies have to learn everything needed for survival, and this takes years. But a computer can be programmed at "birth" with a vast amount of information. Furthermore, intelligent computers could make improvements to their "bodies"—replacing and upgrading parts of them—something humans can't do.

But would a computer civilization want to upgrade itself? Would it "want" to do anything? Could it, indeed, have emotions such as love, hate, greed, happiness, and so on? It seems unlikely, but does a civilization necessarily need these things? I think not.

The second possibility for nonchemical life is the intelligent gas cloud. This idea was used in 1957 by Fred Hoyle in his science fiction book *The Black Cloud*. According to his tale, a giant interstellar cloud that was capable of thought arrived in the vicinity of the sun, its mission to extract energy from the sun. The cloud's thought process consisted of radio

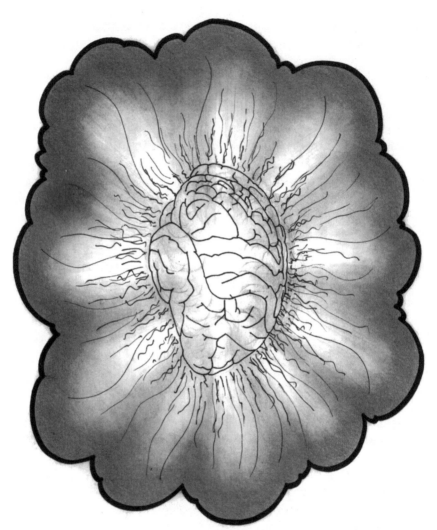

Hoyle's black cloud, a living cloud.

signals from one part of the cloud to another. Some of these signals were picked up by radio operators on Earth, and about the same time the cloud discovered that the Earth was the home of intelligent beings.

Could such a cloud exist? In Hoyle's story the cloud's brain was a large-scale version of the human brain, with the molecules arranged so that neural-like electromagnetic currents passed through the cloud. In theory this is possible, but it would take hundreds of billions of years for such a cloud to form, and the universe is not that old. The problem is twofold: the average density of matter in interstellar space is millions of times less than the density of matter on a planet, and therefore interactions between particles are much slower. The time for life to evolve in a cloud would therefore be millions of times slower. Temperature is also a problem: most interstellar clouds have temperatures that are far below those acceptable for the formation of life. A few, however, might have regions that are reasonable, but high temperatures tend to decrease the density of matter, so there's a catch-22.

Finally, we have the galaxy that is alive, with stars acting as basic units, similar to atoms in matter. This is another intriguing idea; basically, it's life based on gravitation rather than electromagnetic forces. Stars do, indeed, move about within a galaxy, and in theory they could assemble themselves into meaningful units, but again time is the problem. It would take trillions of years for anything like this to happen and the universe is not old enough.

Life at the Extremes on Earth

In the next chapter we will begin looking into the possibility that minor forms of life exist on Mars. We know there are no higher forms of life, but it is possible that elementary forms such as bacteria, fungi, or protozoa might exist there. We have, in fact, found some evidence of this. One of the reasons we expect to find life there is that primitive forms of life are so common on Earth, even under extreme conditions. Marine life and microorganisms exist on Earth at temperatures below freezing, and microorganisms have been found in the Antarctic at temperatures of 15 degrees below zero (centigrade). Furthermore, bacteria and other life forms survive in thermal ponds at temperatures well above boiling. Bacteria and algae live in the radioactive cooling ponds of reactors, virtually immune to the radiation they are absorbing. Bacteria have also been found in the acid of car batteries, and there are bacteria that need no oxygen to survive.

Surprisingly, microorganisms have also been found deep beneath the surface of the Earth. The first evidence of them was discovered in the 1920s by Edson Bastin of the University of Chicago when he examined the water extracted from oil fields. At that time there was a strong possibility of contamination and his suggestion that the microbes were from extreme depths was met with considerable skepticism. There was little further interest in his work until the 1980s, when the EPA became concerned about the condition of groundwater in the United States. Teaming up with the Department of Energy (DOE) and a U.S. Geological Survey group they began looking into the matter. They decided that if there were microorganisms thousands of feet beneath the surface, they could be used to degrade pollutants and radioactive waste.

Because of the difficulty of bringing water from thousands of feet below the surface to the surface without contaminating it, the team decided to search for life forms inside rock. The search began in 1987 beneath the Savannah River in South Carolina. To their surprise they found bacteria were common within the spores of rocks from as deep as 1.7 miles beneath the surface. The temperature at this depth is an incredibly high 75°C (167°F).

Temperature is now believed to be the factor that limits subsurface life. On the basis of known temperature rises beneath the surface we can, in fact, make an estimate of how deep bacteria may exist. Temperature rises at 25°C (45°F) for every 100 meters under the continental crust. If we assume an upper limit for life of 110°C, this means there should be life down to 4000 meters. Under the oceans, however, the temperature rises by only 15°C per 100 meters of depth. Life could therefore exist 7000 meters beneath the ocean floor.

According to the Savannah team the abundance of subsurface microbes varies considerably from place to place. Samples from 400 meters contain from 100 to 1 million bacteria per gram. By comparison, topsoil contains about a billion per gram.

One of the important questions in relation to bacteria this far below the surface is, how long have they been there? The team showed that water from the region had been buried for thousands of years, and in some cases there were indications it had been there for millions of years. The sediments at this level were laid down about 300 million years ago, and if the bacteria are, indeed, this old we wonder how they managed to survive. In some cases essential nutrients are renewed, but this isn't true in all cases. Some of the bacteria have to come close to starving, but they appear to have a mechanism to cope with it. During these starvation periods they shrink up to $\frac{1}{1000}$ their normal volume and lower their metabolic rate by a factor of a million or more. When these same bacteria are on the surface they reproduce in minutes or hours, but beneath the surface they reproduce only once in 100 years or so.

With life able to adapt to such extremes on Earth, it gives us hope of finding life under the extreme conditions that exist elsewhere in the solar system, and in the universe.

Gaia

The concept of Gaia has taken on considerable importance in recent years. To some, in fact, it's a form of life. The word Gaia refers to the total system of life on Earth, in other words, all life forms and everything else related to life. Our atmosphere plays a central role in sustaining the life on our planet. Life began in the atmosphere, and once developed was sustained by the atmosphere, and the relationship between our atmosphere and life is intricate and complex. There's no question that the atmosphere has helped shape life, but what effect does life have on the atmosphere? The British researchers James Lovelock and Lynn Margulis have suggested that life on Earth regulates and controls the atmosphere. It does this by adjusting the amount of various elements—particularly carbon dioxide—in it, which, in turn, controls the average temperature of the Earth. This is the Gaia hypothesis.

We know that there is a complex relationship between the atmosphere and life, but the suggestion that life controls the atmosphere is relatively radical. Most scientists do not accept it, but it has generated a lot of interest in recent years. As an example of this control, consider the case where the temperature of the Earth suddenly drops. According to the Gaia hypothesis this would adversely affect the growth of various organisms, but the death of large numbers of them would increase the carbon dioxide in the atmosphere, which in turn would increase the greenhouse effect, and the temperatures would soon be back to where they were. There is, indeed, some evidence of this. One of the major problems for scientists is, why was the Earth so warm millions of years ago when the sun was much younger? Astronomers can easily calculate the size and surface temperature of the sun as it evolves, and they know that it was about 30 percent cooler when life first formed on Earth than it is today. This is a considerable difference and, according to many scientists, it should have made the Earth too cool for life to develop. Yet the average temperature was roughly the same then as it is today.

We know why the temperature was relatively high at that time. Volcanic activity was more widespread, and volcanoes give off carbon dioxide, so the greenhouse effect was more extensive. But strangely, as the sun slowly warmed up, the atmosphere gradually changed, keeping the temperature relatively constant. This is what the Gaia hypothesis predicts, namely, life on Earth controls the atmosphere, keeping the temperature constant. It is easy enough to check this idea. We know the Earth has undergone several ice ages, and we could look back in rocks for evidence of an abundance of carbon dioxide after each of these ice ages began. So far we haven't found any evidence of this.

On the basis of this, a number of people have suggested that Gaia acts like a single organism, and exhibits life. Lovelock and Margulis made no such claim, but it is an interesting idea.

Gaia may or may not be right, but there definitely is an important interdependence and balance between life and the atmosphere (and many nonlife processes) on Earth, and we should keep this in mind as we turn to the planet Mars in our search for life beyond Earth.

chapter 5

Mars as an Abode for Life

Mars, as seen through a telescope, is a fascinating sight: a tiny red ball with intricate dark markings, capped with a silver crown of ice. Because it is the only planet that has markings on it that change with the seasons, it was a favorite of early astronomers. Many of them stared at it in awe, wondering if it harbored life. Then with the discovery of fine lines—possibly canals—on its surface, the public became intrigued with it. It is perhaps little wonder that when H. G. Wells's novel *War of the Worlds* was broadcast on the radio in 1938 people panicked and fled into the streets, convinced that we were under attack by Martians.

It soon became clear, however, that Wells's story was only fiction, and that Mars is a vast, icy desert with only a thin atmosphere. Any life on the planet would have to be primitive. Then the first flights to Mars brought further disappointment: the landscape was even more harsh than had been anticipated. It was moonlike, covered with craters.

Early Explorers

Despite the intriguing markings on its surface, Mars was a difficult object to see clearly. It is only half the size of Earth, and both our atmosphere and dust as well as the instability of the Martian atmosphere hinder our view. Furthermore, Mars is only ideally situated for viewing at certain times. They occur when it is directly opposite Earth in its orbit—in other

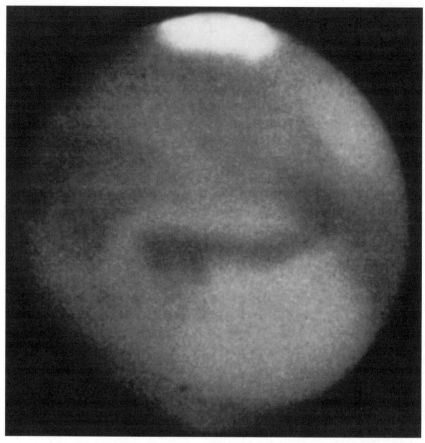

Mars as seen from the Earth through a telescope. (Courtesy NASA)

words, when it is at opposition (when the planets are opposite one another in their orbits). But because of the ellipticity of the two orbits, particularly that of Mars, some oppositions are more favorable than others. At an opposition the two planets can be as close as 35 million miles, or as far away as 63 million miles. Oppositions occur every two years, usually in the months of August or September.

One of the first in a long line of Mars observers was Christian Huygens. Born in Holland in 1629, Huygens was educated at the University of Leiden. His early interest was mathematics—he wrote the first book on probability theory—but he eventually turned to astronomy and physics,

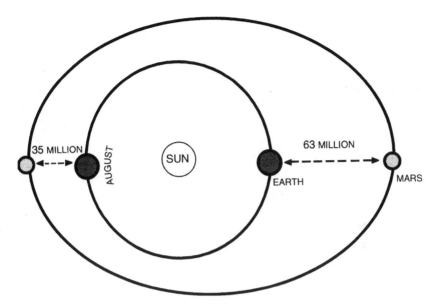

The orbits of Earth and Mars showing oppositions.

and his interest in astronomy led to a new technique for grinding lenses. The lenses he produced were the best in the world, and with them he made several important discoveries. He discovered the large moon of Saturn that is now called Titan, and in 1659—an opposition year for Mars—he turned his telescope to the red planet and discovered a V-shaped marking on its surface, a region we know as Syrtis Major. By watching the dark markings carefully he determined that the planet rotated on its axis in approximately 24 hours.

A few years later a better estimate of its revolutionary period was obtained by Giovanni Cassini, an Italian–French astronomer. Cassini was also able to determine the distance to Mars using a technique called parallax. Parallax is the apparent shift in position that occurs when you observe something from two different positions. You can observe it by holding your finger out in front of you, and blinking your eyes one at a time; your finger will appear to move back and forth relative to the wall behind it.

Cassini's and Huygens's telescopes were of high quality for their day, but vastly inferior to those that were soon to come. Both men used refractors, telescopes similar to the "spotting scopes" of today, where you look directly through the lenses. But refractors had a serious flaw: they produced false colors. We now refer to this flaw as chromatic aberration. After studying it for some time Newton decided the only way around it was to design a new type of telescope. He built a telescope that had no chromatic aberration; it employed a curved mirror rather than a lense, a type we now refer to as a reflector. His telescope, however, was small, its mirror only an inch in diameter. But larger and more powerful reflectors were soon to come.

The man mainly responsible for these larger telescopes was William Herschel. Born in Hanover, Germany, in 1738, Herschel followed in his father's footsteps and joined the band of the Hanover army. He hoped to become a professional musician. But when Hanover was occupied by the invading French army and the band came under fire, Herschel's father, fearing for his son's life, urged him to desert. With help from his family he fled to England.

Although he was almost penniless when he arrived in England, Herschel soon found employment, and within a few years was relatively prosperous. By 1766 he was the organist at Bath, and a well-known music teacher with more students than he could handle. Most of his time was

William Herschel.

taken up with music, but he was an avid reader and eventually came upon books in astronomy and optics. As his interest in this new area increased, music became secondary and he began to spend most of his time studying astronomy. He decided he needed a telescope, but a large refractor was beyond his means, so he decided to build his own. He started with a refractor but soon turned to reflectors. Through trial and error he developed new techniques for casting and grinding large reflecting mirrors. His sister Caroline joined him in England, and she assisted him as he built larger and larger telescopes and began using them to explore the heavens.

In 1781 he discovered the planet Uranus. The discovery brought him considerable fame, and he was given a stipend by the king, which allowed him to spend all of his time observing the sky. He had already observed Mars, and he observed it again at oppositions in 1781 and 1783. He was amazed at its similarity to Earth: it appeared to have seasons as Earth did. Watching it over a period of months, he saw the pole caps recede and a "wave of darkening" spread toward the equatorial regions of the planet. It had to be vegetation. He was soon convinced that Mars was populated by an intelligent race of beings.

Schiaparelli and the "Canali"

Herschel's discoveries created considerable interest, but the observations that had the most influence on the public, and caused the most controversy, came a few years later when faint "lines" were discovered on the surface of the planet. Although they had been seen before Giovanni Schiaparelli first noticed them, it is his name that is most strongly associated with them. Born in Savigliano, Italy, in 1835, Schiaparelli graduated from the University of Turin in 1854 with a degree in architecture and hydraulic engineering. His first job was teaching mathematics in an elementary school, but he soon realized it was a mistake. From early youth he had been fascinated with astronomy, and he realized now that it was his real calling. In 1857 he applied for and received a government stipend for study abroad. He used it to study astronomy at the Royal Observatory in Berlin and at an observatory in Russia, where he worked with the well-known astronomer Wilhelm Struve.

Upon returning to Italy, he obtained a position at the observatory at Brera Palace in Milan. For the first few years he used the observatory's 8.6-inch refractor to study double stars, but in 1877 all eyes were turned to

Giovanni Schiaparelli.

Mars. It was an opposition year, one of the best in decades; Mars would be only 35 million miles away. Schiaparelli studied the planet, making careful micrometer measurements of markings on its surface. He had a sharp eye, a telescope power of 322, and many clear evenings of excellent seeing. Waiting patiently for short periods when both the atmospheres of Earth and Mars were still, he was able to get glimpses of unprecedented clarity. It was during these periods that he began to make out faint lines on the surface. Only two or three appeared at a time, but he marked their positions carefully on his maps. In the process he drew up a new map of the planet, renaming most of the features, discarding old names such as Kaiser Sea, Herschel Strait, Hooke Sea, and Cassini Land, and replacing them with Syrtis Major, Sinus Sabaeus, Tyrrhenian, and Ausonia. At first his new nomenclature was viewed with skepticism by astronomers who were intimately familiar with the old names, but as his maps became better known most astronomers realized his names were more appropriate than the old ones, and they were adopted.

To Schiaparelli the lines were a natural phenomenon, and he referred to them as *canali*—Italian for channel. But in the translation to English the

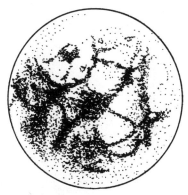

A drawing of Mars by Schiaparelli, showing "canals."

i was lost, and they were taken to be canals. Canals are, of course, the product of a civilization, and within a short time he and his discovery were making headlines.

Was it possible that Mars harbored a civilization? The indications were strong, but most astronomers rejected the idea. Little was known about the planet; it had an atmosphere, but its composition was still unknown. Schiaparelli, although he was largely responsible for the "canal fever," never seriously believed they were made by a race of civilized beings. For the most part he stayed out of the controversy. There were, however, fervent believers in the idea. Camille Flammarion, of France, was one. Best known for his book *The Planet Mars*—a compendium of all the important studies that had been made of the planet—he was sure Mars was inhabited, and speculated that the Martian civilization was far more intelligent and technologically advanced than ours.

Lowell

For the next few decades the Martian scene was dominated by one man—Percival Lowell. He became the leading authority on the planet, setting up an observatory in Arizona that was devoted almost entirely to its study. Born in 1855 in Boston to an aristocratic family of achievers—his sister Amy was a well-known poet, and his brother became the president of Harvard University—Lowell graduated from Harvard with honors in 1876. His majors were English and mathematics, but even then he was

Percival Lowell.

dabbling in astronomy. At his commencement, he gave a short talk on Laplace's theory of the origin of the solar system. For the next few years he worked in his grandfather's textile mill but eventually got itchy feet and decided to travel. For years he had had a serious interest in the Far East, and in 1883 he left for the first of three trips to that region. He studied the cultures of both Japan and Korea and wrote several books on them.

Upon returning to the United States in 1893, he read Flammarion's book on Mars and was intrigued. Within a short time he was corresponding with William Pickering, brother of the director of the Harvard Observatory. Pickering was looking for a donor to fund an expedition to the Arizona desert to observe Mars during the upcoming opposition. Lowell offered to fund the project, but to Pickering's surprise he also wanted to take part in it.

Lowell, Pickering, and Andrew Douglass, a young graduate from Connecticut, arrived in Flagstaff, Arizona, in May 1874 and had soon set up a makeshift observatory of tents. They had two telescopes with them: a 12-inch and an 18-inch refractor.

Flagstaff, at an elevation of 7000 feet, was surrounded by desert, so the air was clear and the "seeing" (the steadiness of the atmosphere) excellent. After barely a month of observation, Lowell began writing articles about the planet, not for scholarly publications but for *Popular Astronomy* and *The Atlantic Monthly*. He returned to Boston shortly thereafter and gave a series of talks on the planet; he also began preparing for the publication of a book on the planet that was later titled *Mars*.

Lowell was convinced that Mars was a dying planet inhabited by a race of beings that had built a gigantic system of canals to conserve the last of their water. According to his ideas, as the pole caps melted (he assumed they were water ice), the water was brought via the canals to the equatorial regions where it was used to irrigate crops.

The public was soon caught up in Lowell's speculations, and he became a well-known figure and authority on the planet. Astronomers, on the other hand, were skeptical of him, but once his observatory was established and the output from it was shown to be of high quality, they realized that his contributions were important.

Some astronomers saw Lowell's canals; many, however, were convinced that he was seeing things and had an overactive imagination. The debate raged for years. Then the large 100-inch Hooker reflector went into operation at Mt. Wilson in California, and it was soon turned to the red planet. Because the new reflector was much larger than any of the telescopes at Lowell's observatory, and was under equally stable skies, astronomers expected it to give a greatly enhanced image. But, strangely, the image was similar to the one seen at Flagstaff; the photographs were no better. Canals did not show up on them. But with the larger objective—the 100-inch mirror—something new was possible. There was now enough light to get a good spectrum of the planet, and it gave some interesting surprises: Mars was quite different from Earth. Its atmosphere was mostly carbon dioxide, and the temperatures on the planet were extremely low. Although they could be relatively high (60°F) when the sun was overhead in the equatorial regions, they plunged to −123°F at night. There was little chance that a higher form of life could exist there. But there was still the hope that a lower form of life might exist.

Mariner Flights

The controversy over life was far from over when the first space vehicle to Mars, *Mariner 4*, was launched in November 1964. Astronomers were not sure what to expect as the spaceship neared Mars, but anticipation was high. The first images were fuzzy but they left little doubt: the surface was heavily cratered. It looked like our moon. This was a surprise, and a disappointment to most. Mars was, after all, quite different from our moon: it had an atmosphere, and even though it was thin compared to Earth's atmosphere, it should have had an effect on craters. They should have been eroded as they are on Earth. Furthermore, it was known that huge dust storms raked the surface, sometimes covering a large fraction of it. It seemed strange that they had done little to hasten the process of erosion.

Although the craters were shallower than those on the moon, they were just as numerous and had been there for several billion years. Furthermore, the atmosphere was verified to contain mostly carbon dioxide—95 percent of it was made up of it—but the pressure was only $\frac{1}{100}$ that of the Earth's atmosphere. This was a surprise. With such a low pressure, liquid water could not exist on the surface, and without liquid water, vegetation—even primitive forms such as lichen—could not exist.

But *Mariner 4* saw only a small fraction of the planet—about 1 percent. What about the rest of the planet? Was it cratered in the same way? To find out *Mariner 6* and *Mariner 7* were launched in 1969. This time about 10 percent of the surface was seen, and there was more disappointment. It was cratered as extensively as the section seen by *Mariner 4*.

In 1971 a more ambitious mission was launched; called *Mariner 9*, it would go into orbit around Mars, and over several months, photograph its entire surface. To the dismay of the *Mariner 9* scientists, however, a planet-wide dust storm greeted them when the craft reached Mars.

Mariner 9 went into orbit around Mars and while it waited for the dust to settle it aimed its cameras at Mars's two tiny moons, Deimos and Phobos. They were only about 10 and 15 miles across, respectively, and both were heavily cratered.

In late January 1972, the storm began to dissipate and several circles appeared above the dust. As the dust settled, the tops of four volcanoes were seen. The largest, now called Olympus Mons, is 310 miles across at the base, and has a caldera (crater at the summit) 40 miles in diameter at its peak. Its height, 15 miles, is twice that of Mt. Everest (measured from sea level).

Then came another surprise. The planet was not as dead as expected. The northern hemisphere of Mars was quite different from the southern one, the one that had been photographed on the earlier *Mariner* missions. It not only contained volcanoes but was generally much younger than the heavily cratered southern hemisphere; furthermore, there was considerable evidence of lava flow.

The highest volcano on Mars—Olympic Mons. (Courtesy NASA)

But why are all the volcanoes on Mars so much larger than those on Earth? After all, Earth is a much larger planet. The reason is that the Earth's geology is dominated by plate tectonics (regions or "plates" on the surface of the Earth drifting relative to one another). Volcanoes on Earth can grow for only a few thousand years before the plates on which they reside move away from the source of the magma (molten rock), and they become extinct. (This is seen in the Hawaiian chain; only the volcanoes on the large island at the end of the chain are still active.) On Mars, with little or no plate tectonics, volcanoes remain indefinitely over their source of magma and continue to grow as long as magma is available. Furthermore, Mars's crust is thicker than Earth's, and it can support a greater weight, so volcanoes can grow larger. And finally, Mars's gravity is considerably less than Earth's. Still, there is a limit, and we are no doubt seeing it in Olympus Mons.

The region the large volcanoes are located on was also a surprise. Called Tharsis Ridge, it bulges out from the normal surface by about six miles. Scientists have not yet been able to explain this bulge.

Another of the amazing discoveries of *Mariner 9* was a gigantic canyon, now called Valles Marineris, that runs along Mars's equator. It is 3100 miles long, and if placed in the United States would extend from one coast to the other. Our Grand Canyon is tiny by comparison. What caused such a large canyon? Scientists are still uncertain, but it may have been an early stage of plate tectonics—a breakup of the crust into plates that barely got started.

Of all the discoveries, however, the most startling was what appeared to be old riverbeds. They are channels that have the characteristics of riverbeds: shape, sandbars, an apparent downhill flow, and in many, a dendritic appearance typical of river systems on Earth. For those searching for evidence of life the fact that there was water on the surface at one time would have tremendous implications.

But what about the canals? Strangely, there was no sign of them. Astronomers now believe that they were chance alignments of objects on the surface that gave the appearance of lines. In some cases they may have been optical illusions. Also, there is the "wave of darkening" seen each spring. Early on it had been assumed to be a form of vegetation—perhaps lichen. There was, of course, no sign of vegetation, and it is now assumed that this wave of darkening is due to the dust that is blown into the atmosphere each spring. As it is lifted into the air, dark regions beneath it are exposed. (All aspects of this wave are not explained this simply, however, and there is still some controversy.)

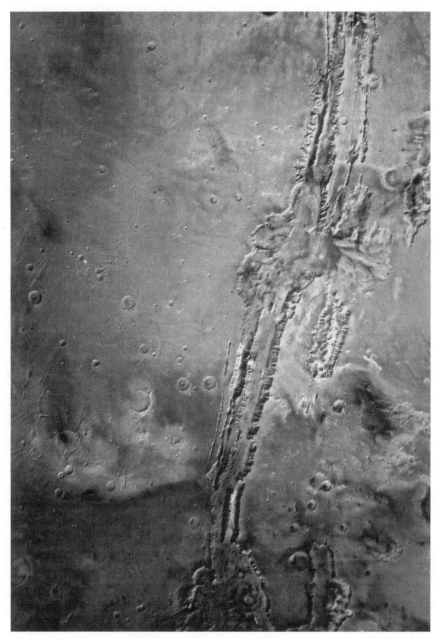

The "Grand Canyon" of Mars, Valles Marineris, or the Valley of Mariner. (Courtesy NASA)

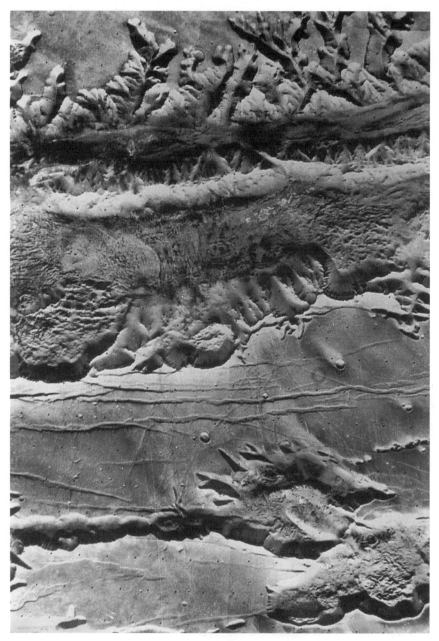

Close-up of the Valley of Mariner. (Courtesy NASA)

Mariner 9 brought many pleasant surprises, but most of all it brought a renewed hope for finding life. With the earlier flights hope was lost, but now that evidence of ancient riverbeds had been found, there was the chance that at least a primitive form of life existed. A landing was therefore needed to check the soil.

Viking

On July 20, 1976, exactly seven years after the first manned landing on the moon, *Viking I* landed on Mars. The first of two spacecraft, it landed in the Chryse basin in an area that was relatively flat. No mountains were visible, but small sand dunes could be seen in the distance, and on all sides were fields of boulders. Beneath the boulders was a fine, flourlike sand; both sand and boulders were red.

To the surprise of many the Martian sky was not blue as it is on Earth. Because of the red sand that circulates continuously in the atmosphere, it was pink. And the atmosphere itself was found to have the composition 95 percent carbon dioxide, 2 percent to 3 percent nitrogen, and 1 percent to 2 percent argon, with trace amounts of oxygen and water vapor.

The second craft, *Viking II*, landed on the Plain of Utopia a month later. The view was about the same, with large fields of boulders in all directions. If anything, it was even flatter than the region around *Viking I*.

While the landers checked the surface, the two orbiters photographed the surface from above. One region of particular interest was the polar caps. For years astronomers had believed they were frozen carbon dioxide. As summer approached in the northern hemisphere, however, the north pole cap melted and receded, but strangely, even when the temperature was far above the sublimation point for dry ice (solid carbon dioxide), a portion of the cap remained. This meant that it had to be water ice. The upper layer of the cap was apparently carbon dioxide ice, but when it sublimated it revealed a layer of water ice beneath it.

There was also other evidence for water ice on the planet. The chaotic terrain at the end of Valles Marineris appeared to be a region that had sunk or collapsed when ice beneath the surface melted and the water flowed out. And as we will see later, the appearance of many of the craters on the planet indicated that there is frozen water beneath the surface.

Arm of Viking craft digging a trench. (Courtesy NASA)

Viking lander. (Courtesy NASA)

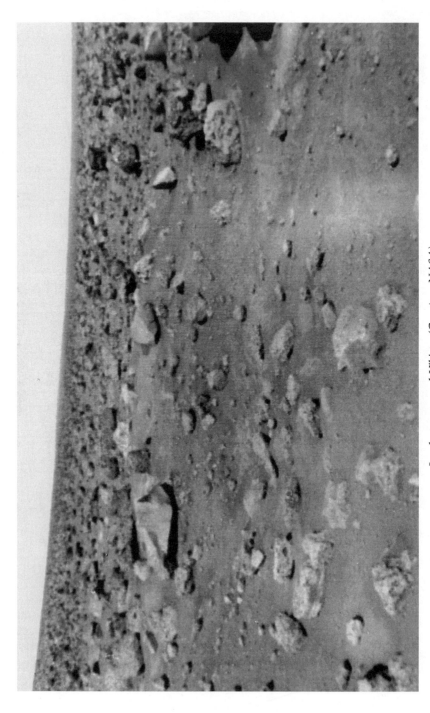

Landscape around Viking. (Courtesy NASA)

The Oceans of Mars

Although there are obviously no higher forms of life on Mars, it is in many ways more intriguing than anticipated, and we may learn much about the Earth's history by studying it because its early history was similar to Earth's in many ways.

But of most importance, there is now no doubt that there was once liquid water on the surface. Some of the old riverbeds are gigantic by Earth's standards—up to 50 miles across. (Most rivers in the United States are less than a mile across.) Today water cannot flow on the Martian surface because of the low atmospheric pressure, but in the past Mars must have had a much denser atmosphere.

The atmosphere of Mars, like that of Earth, was expelled from volcanoes early in its history. Light gases such as hydrogen easily escape to space on Earth. On Mars, with its weaker gravitational field, heavier gases such as nitrogen also escape. A heavy gas such as argon, however, is retained almost indefinitely. Almost all of the argon that was ever expelled from volcanoes on Mars is therefore still in its atmosphere. Since volcanoes release a known fixed ratio of gases, by measuring the amount of argon now present, we can determine how much nitrogen, carbon dioxide, and so on was expelled in the past. The amount of argon in the Martian atmosphere is about $\frac{1}{25}$ that of Earth's atmosphere, indicating that a much higher density of nitrogen and carbon dioxide was present in the past. Calculations indicate it would easily have been enough for water to flow on the surface.

Evidence from oxygen abundance supports this. There is only a tiny amount of oxygen in the Martian atmosphere—less than 1 percent compared to Earth's 21 percent—but it has been particularly helpful in giving us information. Most oxygen is what is referred to as oxygen-16, but there is an isotope with fewer particles in the nucleus called oxygen-14. Oxygen-16 is slightly heavier than oxygen-14, and is therefore lost to space at a slightly slower rate. By comparing the present ratio of these isotopes it is possible to estimate how much oxygen was originally in the atmosphere. Calculations indicate that the overall pressure may have been twice that of the Earth's atmosphere.

Furthermore, if there once was a considerable amount of oxygen in the atmosphere (most of which is now locked up on the surface in the form of iron oxide) we have to ask where it came from. On Earth all the oxygen came via photosynthesis after vegetation appeared on the surface (and algae in the oceans). Was there vegetation on the surface of Mars to supply the oxygen? Scientists know of no other way to explain it.

But if the atmosphere was denser in an earlier era, the temperature on the surface also had to be higher. If the atmosphere had more carbon dioxide in the past, for example, the greenhouse effect would have been greater, and it would have raised temperatures. According to recent calculations the greenhouse effect could have raised the average temperature as much as 60 degrees. And with all the water now in the polar caps and trapped beneath the surface as permafrost, if the atmosphere was denser this water would have flooded onto the surface. Indeed, there is enough ice on the planet now that if it was all in the form of water would cover the planet to a depth of 30 feet (if uniformly distributed). Furthermore, there was likely more water in the past than there is today.

But if there was a lot of water on the surface in the past, where was it located? Water on the Earth eventually flows into oceans. Was this the case on Mars? Many scientists are convinced that it is. The evidence, in fact, is overwhelming. Many of the large channels, for example, flow toward lowlands, regions that are surrounded by steep cliffs, reminiscent of our own continental shelves (the shallow regions surrounding the continents). Further evidence comes from the craters. An examination of the region around many of them shows that a slurry of mud appears to have flown out when the asteroid that created them struck. The impact apparently melted ice beneath the surface and threw it outward as mud. Interestingly, though, all craters on Mars are not of this type; most of the craters of the highlands are not. Also, in the equatorial regions, large craters (over three miles across) have muddy ejecta around them, but smaller craters do not. This indicates that the icy region is deeper near the equator.

Of particular importance, though, there are mud slurries around most craters over a large region where an ocean is believed to have existed. This ocean is now assumed to have covered much of the northern hemisphere; it has been named Oceanus Boreallis and may have been four times as large as the Arctic Ocean on Earth. Smaller bodies of water are also believed to have resided in other regions.

A number of scientists have suggested that the water of this large ocean eventually sunk into the soil and froze, creating the large sheets of underground ice that we have detected. Both *Viking* landers touched down in regions that are believed to have at one time harbored the ocean, and tests of the soil indicate that this is the case. The soil has a large amount of iron oxide in it—in particular, a form called limonite. Limonite is formed on Earth only under conditions of high humidity where there is a good supply of oxygen and iron.

Water on early Mars.

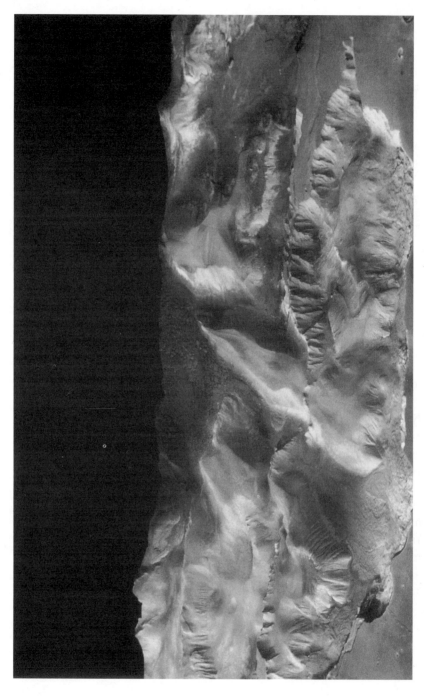

Mountains on Mars. (Courtesy NASA)

Ice Ages on Mars

But if Mars once had an almost tropical climate and water on its surface, how did it end up as the frozen wasteland we now see? For an answer we need only look to Earth. We know the Earth's climate has undergone many changes in the past due to ice ages, the last *as recently as 10,000 years ago*.

A possible explanation of these ice ages was given by the Yugoslavian scientist Milutin Milankovitch in the 1950s, but at the time few took it seriously. Many of his predictions, however, have turned out to be correct, and his idea is now believed to be true. Milankovitch proposed that ice ages are caused by three different phenomena:

1. Oscillations in the shape of the Earth's orbit.
2. The varying tilt of the Earth's axis. It is presently 23.5 degrees, but varies from 22 to 25 degrees.
3. The precession of the Earth's axis. This is the conical motion that the axis of any spinning object (e.g., a spinning top) makes. In the case of the Earth it takes approximately 19,000 years to trace out the cone.

Milankovitch predicted the dates of several of the recent ice ages, and evidence from ocean bottoms indicates he was correct.

If we look at the same effects on Mars we see that they are considerably greater and would cause more intense ice ages than on Earth. First, the eccentricity of Mars's orbit is much larger than Earth's; at one end of its orbit it is 30 million miles closer to the sun than at the other end. The Earth's distance from the sun varies by less than a million miles. Also, as has recently been discovered by Jack Wisdom and Jihad Touma of the Massachusetts Institute of Technology (MIT) the tilt of Mars's orbit varies considerably more than Earth's: from 11 degrees to 49 degrees over a period of 3 to 4 million years. This would have a dramatic effect on the climate.

The ice cap at Mars's north pole has a layered structure that is also indicative of ice ages. It appears to have been caused by thawing and freezing over millions of years. There is also evidence of large glaciers on Mars in the past. Bouldery ridges of sediment left by melting glaciers and flow lines characteristic of glacier flow are seen in many places.

Besides this, it is well-known that our sun goes through cycles where there are few sunspots, and they cause mini ice ages. Two recent eras of

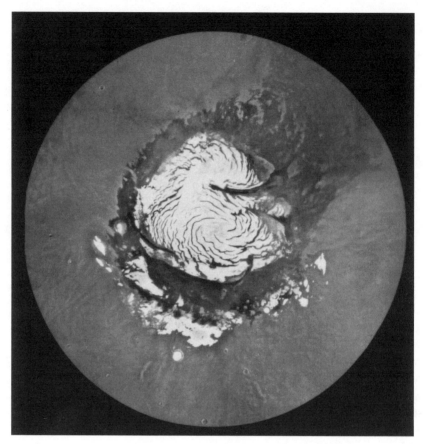

Close-up of north pole ice cap on Mars. (Courtesy NASA)

this type are well-known: the Maunder minimum and the Spörer mini-mum. Winter temperatures on Earth were slightly lower during these minima, but in the distant past these eras may have been more severe, and they may have had a serious effect on Mars.

One of the difficulties in relation to life on Mars is that according to most statistics it has been almost a billion years since water flowed on its surface. A study of the riverbeds by Harold Masursky of the U.S. Geological Survey, however, has shown that there were likely several eras of rainfall and flooding. He examined the density of craters on the riverbeds and found that they differed significantly. Some had few craters, others a large number. According to his estimates these eras of rainfall spread from 3.5 billion years ago to less than half a billion years ago.

Looking for Life: Viking Results

Because of the strong evidence that Mars once had an atmosphere, surface water, and moderate temperatures, it is reasonable to assume that life may have developed, and if so, there may still be a lower form of life in the soil. The major objective of *Viking* was to establish if microbes were, indeed, present. The soil is the logical place to look because Mars has a thin atmosphere with practically no ozone in it, and ultraviolet radiation from the sun—detrimental to life—reaches its surface. Any life forms would need protection, and the only protection would be the soil.

It might seem that it would be simple to test for life, but this isn't so. Life is complicated, and identification of microbial life is not an easy task. It can be seen with a powerful microscope, but weight limitations would not allow the transport of such an instrument to Mars. Scientists knew it was going to be difficult to identify life, and as it turned out it was even more difficult than they anticipated. There were a total of five tests. The first was the camera view around the lander, and of course it turned out negative. A check of the soil was made for dead organic matter and none was found. And finally there were three biological experiments, each quite sophisticated.

To understand the biology experiments let's consider microbes here on Earth. We know that they take in various substances, break them down rearranging them chemically, and then release by-products, usually gases. The three biological experiments aboard *Viking* attempted to identify microbes by trying to initiate this cycle, for example by feeding the microbes a nutrient and looking for a gas that was given off.

In the first experiment, the pyrolitic release experiment, a spoonful of Martian soil was put in the test chamber along with a simulated Martian atmosphere. A small amount of radiaoactive-labeled carbon dioxide was added to the atmosphere so it could be identified later if it showed up in the products of the reaction. (The radioactive carbon dioxide did nothing to alter the results.)

When the chamber was closed a lamp was turned on to simulate Martian sunlight. Algae and other plant life in the soil would, in theory, undergo photosynthesis; in other words, they would absorb carbon dioxide and convert it to organic molecules. After a five-day incubation period, the chamber was heated to 1200°F to break down and vaporize the cells of any organisms present. The resulting gas was then fed to a gas chromatograph for analysis.

To their surprise scientists detected considerable radiation, which indicated that photosynthesis had indeed taken place. Before it could be accepted as a positive result, however, it had to be verified. In particular they had to be sure the same reaction didn't occur in sterilized soil, and that it was not due to a chemical reaction that didn't involve living organisms. To do this the Martian soil was sterilized by heating it to 340°F for three hours; this would kill anything living in it. The experiment was then done again. No radiation was detected in the second test; any organisms in the soil had apparently been killed. Then to be doubly sure, the experiment was run with the lights turned off; photosynthesis would presumably not take place if there was no light. Again, no radiation was detected.

Strangely, though, the *Viking* team was hesitant to interpret this as a positive result. It seemed more likely to them that an unanticipated chemical reaction had occurred, and the result was due to chemical reactions, not biological ones. They knew that the suggestion that there was life on Mars, even microbial life, would cause a tremendous amount of excitement, and because of this the *Viking* team remained cautious.

The second experiment, referred to as the labeled release experiment, also gave positive results. In this experiment a nutrient solution consisting of sugars that had been radioactively labeled were added to the soil. Any microbes in the soil would presumably take in the nutrient and release gas. And indeed, there was a massive outpouring of radioactive gas. Again, to be sure it wasn't just a simple chemical reaction, the soil was sterilized and the experiment was repeated, and no gas was detected. Sterilization was also done at a lower temperature of 122°F, low enough so that it could not alter any of the chemicals, and again the result was negative.

The third experiment, the gas exchange experiment, was similar to the second. In this case a nutrient containing carbohydrates, protein, and vitamins was added to the soil; the team members jokingly referred to it as "chicken soup." After the soup was added, scientists monitored the air in the chamber above the soil for changes. In this case there was also an outpouring of gas, namely oxygen, from the soil. It was interpreted as a positive result, but strangely the reaction did not continue to increase as expected; within two days it had leveled off. The scientists on the team were not able to explain this biologically, so they eventually decided that what they were seeing was an unexpected chemical reaction. But again the reaction was eliminated by sterilizing the soil, as it would be if it were caused by microbes.

After careful study the team decided that they could not say with any degree of certainty that they had detected life in any of the experiments. (There was, however, some disagreement. One team member, biologist Gilbert Levin, is still convinced that the results indicated microbial life.) One of the major reasons for this consensus was the result from the gas chromatograph mass spectrometer (GCMS). The GCMS was designed to look for organic material in the soil, in other words, the remains of dead organisms. It wasn't looking for living organisms and, even if they were there, it would not have been able to detect them, but it was particularly sensitive to organic material, and it detected none. This was a surprise, considering the results of the biological experiments. But if there *were* live microbes on Mars, where were their dead bodies? They couldn't just disappear. One possibility is that the live ones are scavengers and eat organic material. At any rate, the result was strange, and it cast a shadow on the two biological experiments that appeared to give positive results.

Several scientists have pointed out that we do not know enough about life outside Earth to rule out life on the basis of these experiments. And as we saw earlier, at least one of the team members was convinced that life was detected. We will look into this further in the next chapter.

chapter 6

Meteorite from Mars

Four and a half billion years ago, as the crust of Mars was cooling, a small section of lava a half mile beneath the surface solidified. About the size of a potato, it became igneous rock. To most it would have looked like an ordinary rock, but this small section of crust would eventually generate tremendous excitement here on Earth. It would become meteorite ALH84001.

Just over a billion years later a huge irregular chunk of rock—an asteroid—struck the Martian surface nearby, sending out shock waves that passed through ALH84001, creating fissures throughout it. Mars was in a state of turmoil at this time, its surface, covered with volcanoes, spewed out carbon dioxide and other gases. The atmosphere was much thicker than it is now and there was water on the surface. Some of this water eventually trickled down to the small section of rock that would become meteorite ALH84001. The water crept into the cracks and fissures caused by the nearby impact, and in this water were microorganisms, so tiny that millions would fit on the head of a pin. Nourished by the water, they grew and reproduced, but eventually the water dried up and they died. Finally they fossilized, turning to stone.

The small piece of rock lay beneath the surface undisturbed for another 3.6 billion years. But turmoil continued in the skies over Mars. Asteroids continued to crash into Mars's surface. One of these asteroids struck the surface at a sharp angle. It shattered the rocky surface, digging a deep elongated trench, throwing pieces of rock out at velocities five times faster than a rifle bullet. At this speed many of them escaped the gravitational pull of Mars and ended up in orbit around the sun. The meteorite that would eventually become ALH84001 was among these rocks.

Where on Mars did this asteroid strike? Dr. Nadine Barlow of the University of Central Florida believes she knows. As part of her graduate work at the University of Arizona in the mid-1980s, she compiled a catalogue of 42,283 craters of Mars, and became intimately familiar with them. Working from what she knew, she studied each of the craters carefully. The meteorite had to have come from the oldest part of Mars, the most ancient terrain, but the crater itself, which was made only 16 million years ago, had to be young. It also had to be in a region where there had been considerable water in the past. Because it had thrown the material to space, the imprint of the asteroid had to be elongated, and finally, the crater had to be near another even larger crater, the one that caused fissures in it.

Barlow narrowed her search down to 23 possible craters. Then, using high-resolution photos from *Viking* she eliminated all those with an age greater than 16 million years. When she was finished only two candidates remained. She is still uncertain which of the two is the best candidate but is confident it has to be one of them. One is in the Schiaparelli Impact Basin, the other on the Plains of Hesperia.

The Continuing Journey

As the 4.2-pound rock floated through space it was bombarded by cosmic rays, and these rays changed it chemically, producing slight variations in the elements (new isotopes). Scientists were able to measure these changes and from them they determined how long it had been in space. They believe it orbited the sun for approximately 16 million years.

Most asteroids and planets in the solar system reside in stable orbits, and remain in them for billions of years. But when material from a planet is hurled into space, it can end up in a chaotic orbit. Orbits of this type change significantly over time. Many of the objects in the early universe were in chaotic orbits; this is why there were so many collisions. Most of these "chaotic objects" eventually collided, however, and disappeared from the solar system. The piece of rock that became meteorite ALH84001 was in a chaotic orbit, and over time its orbit took it close to Earth. According to computer simulations, about 4 percent of the rocks blasted into space from Mars eventually approach Earth. About 13,000 years ago the Martian rock came under the influence of the Earth's gravitational field, and it crashed

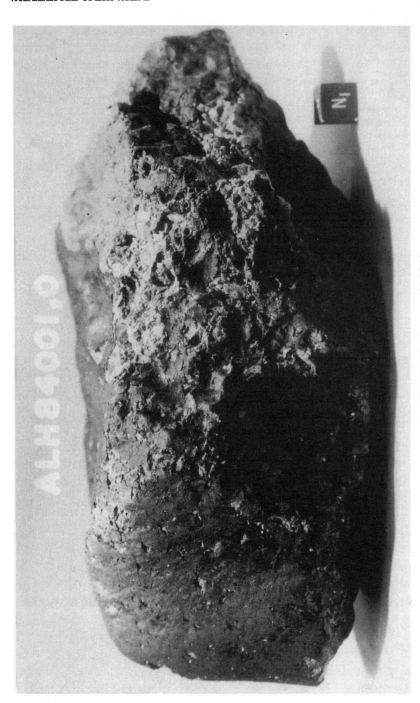

Meteorite ALH84001. (Courtesy NASA)

into the ice of the Antarctic. In its journey down through our atmosphere it was heated by friction from the air, and when it landed it was hot. We cannot be sure how far it penetrated the ice, but it was soon covered with ice and snow, and spent most of its time beneath the ice. But the ice flows collided and reacted, and eventually ALH84001 was brought to the surface.

Other Visitors from Space

ALH84001 is not our only visitor from Mars. Twelve such meteorites have now been identified, but strangely a little over a decade ago we had no idea they were from Mars. In 1980, however, one called EETA97001 was discovered in Antarctica. When examined, it was found to contain regions of glass inside it, and lo and behold, inside the glass were tiny bubbles of gas. Scientists measured the composition of the gas bubbles and were amazed; the mixture was strangely familiar. Just a few years earlier the same mixture had been measured in the atmosphere of Mars. With the landing of *Viking* in 1976 scientists were able to measure the amount of carbon dioxide, nitrogen, and oxygen in the Martian atmosphere, and even more important, they were able to measure the amounts of the rare gases xenon, krypton, neon, and argon in it. Furthermore, it was known that when a meteorite is blasted from the surface of a planet it is likely to trap some of the planet's atmosphere in it. The gases in the bubbles of EETA97001 were *an exact chemical match* to the Martian atmosphere. The meteorite was from Mars.

Scientists were overjoyed. They had a sample from Mars without the expense of launching a rocket to the planet. Furthermore, within a short time, other meteorites were found with the same distribution of gases in them. By 1993 a total of 12 were known. All 12 have a gas distribution identical to that of the Martian atmosphere. No other meteorites have this distribution of gases in them. Two isotopes of oxygen tie the knot particularly tight. The oxygen you breathe is primarily oxygen-16, but there are two heavier isotopes called oxygen-17 and oxygen-18. The ratio between the amount of these two isotopes is well-known, both on Earth and on Mars. The match to Mars is exact.

Six of the twelve Martian meteorites were found in Antarctica; the others are from several different countries around the globe. The first two found were picked up in the 1800s; one was found in Chassigny, France, and the other in Shergotty, India. A third was found in 1911 in Nakhala,

Egypt. Scientists now refer to the group after the first letters of these three locations; they are called the SNCs, or "snicks."

All 12 of these meteorites have been dated, and all except ALH84001 have ages less than 1.3 billion years. ALH84001's age (the time since it solidified) is 4.5 billion years. Early tests on the meteorites indicated that they could not have come from a small body such as an asteroid, as many meteorites do. They had to come from a planet. But it wasn't until after 1980 that scientists knew they were from Mars.

Finding ALH84001

The meteorite ALH84001 sat under the ice in the Antarctic for 13,000 years. Gradually, however, ice upheavals shoved it to the surface. It is, perhaps, remarkable that it was found at all. Why, in fact, would geologists and astronomers ever look for meteorites in Antarctica? It may seem like a strange place to look when meteorites fall randomly over our entire globe; just as many fall in North America in a given area as in Antarctica. If you think about it though, there's an obvious reason to look in Antarctica. The hills and mountains of America are littered with rocks; even when you see a meteorite fall it's difficult to find, mostly because it's so hard to distinguish it from the other rocks that litter the countryside. In Antarctica there are thousands of square miles of ice—nothing but ice—and any rock found lying on it must have come from above, from space. The nearest terrestrial rock is several thousand feet down beneath the ice.

Any rocks found lying on the ice are therefore meteorites, and over the past few years several thousand have been found. Teams go to the Antarctic in search of these meteorites. Roberta Score was on one of these teams in 1984. She and several colleagues set out over the ice one day in December on snowmobiles to hunt for meteorites. Having found many earlier, they knew where to look. They headed for the Allan Hills ice fields of eastern Antarctica.

"We were just cruising around having fun," said Score. They were in a region about 45 miles from the main Allen Hills ice field when suddenly they saw ice pinnacles jutting out of the flat, featureless landscape. Score became concerned, knowing that there could be crevasses in the region, so she proceeded cautiously. As she moved forward on her snowmobile, she spotted the softball-sized meteorite sitting in the ice. She knew immediately it was a meteorite, but it was different from the others she had

Roberta Score.

found, and she knew it would be interesting. It appeared green to her in the bright Antarctic sun, but later she saw it was gray. "It stood out in my mind as being kind of weird," she said. But she had no idea how important it would eventually become.

The four-pound meteorite was shipped to Johnson Space Center (JSC) in Houston, Texas, where it was classified as a diogenite, a relatively common meteorite, believed to be chips from the asteroid Vesta. The rock sat on the shelf until 1993 when David Mittlefehldt, an expert on diogenites, decided to test it. He was surprised to find that it wasn't a diogenite, but had characteristics similar to the meteorites that had been identified as being from Mars.

Within a short time scientists knew that they had found another meteorite from Mars.

The Study Begins

One of those who became fascinated with the rock was David McKay, a geologist at JSC. He began studying the rock in earnest in 1994. If it was

David McKay.

indeed from Mars and was 4.5 billion years old, he was sure it could tell him something about the early history of Mars. McKay also knew that during an early era Mars had water on its surface, so in the back of his mind there was also the possibility that it might tell us something about life forms that may have been present at the time.

Slicing deep into the meteorite, McKay and his team found that it was composed mainly of pyroxene, a silicate mineral that contains magnesium, small amounts of iron, aluminum, and calcium. They also found tiny orange–brown globules lying in the cracks within the meteorite. They were identified as carbonate, a mineral that is formed when water is present. But of particular importance, carbon in any form is suggestive of life, since life on Earth is based on carbon. Furthermore, the globules were surrounded by dark rings. (The rings were found in a sample from the rock about the same time by David Mittlefehldt, the diogenite expert.)

How old were the globules? When were they formed? Tests soon showed that they were 3.6 billion years old, which meant that they were slightly over a billion years younger than the meteorite itself. Why so much younger? The most reasonable explanation is that 3.6 billion years ago water trickled in through the fissures in the meteorite and the tiny globules of carbonate were deposited by the water.

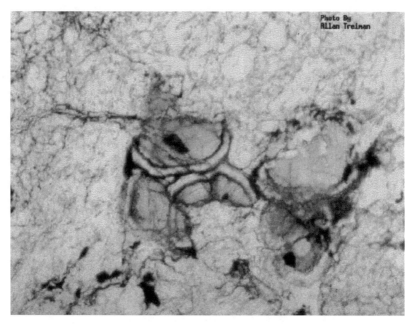

One of the globules in ALH84001.

Others were soon attracted to the project. Chris Romanek, a geochemist at JSC, became intrigued with the tiny globules, as did Everett Gibson, Jr., a senior NASA scientist. Romanek and Gibson looked into the temperature at which the globules had been formed and found it to be relatively low, between 32°F and 176°F. This was encouraging news, because it's the temperature range in which life thrives.

In the meantime McKay began using a powerful scanning electron microscope (SEM) to study the fine structure of the carbonate globules. If life had resided in the tiny cracks and crevices in the rock several billion years ago, there might be evidence of it in the form of microfossils. After all, microfossils of tiny microbes that lived billions of years ago have been found on Earth. With his electron microscope McKay could magnify them up to 50,000 times.

At this stage McKay's group and the Gibson–Romanek team were working independently, but in the summer of 1994 they joined forces and formed a team of nine members, with McKay as the project leader. McKay soon recruited longtime colleague, Kathie Thomas-Keprta to the project.

She was an expert on the transmission electron microscope (TEM), an instrument that would be extremely helpful in the study. Where the SEM could look only at surfaces, the TEM could probe deeply into the material. McKay assigned her the crucial job of searching for tiny fossils—microbes that had hardened to rock billions of years ago.

Thomas-Keprta worked diligently and obtained excellent photographs of the fine structure of the globules, but she found nothing that pointed directly to life. Still, the group was confident they would find something. They had evidence of water, carbonates, reasonable temperatures, and an age that was consistent with life forms. But they knew they needed a lot more. Life in a meteorite from Mars would be a tremendous discovery, one that would get headlines around the world, and they had to be sure of their facts before they made a public announcement.

McKay decided to get in touch with one of the world's top authorities on microfossils, William Schopf of UCLA. Schopf had identified the oldest fossils on Earth, and he could no doubt give them some good advice. Although he was intrigued with their evidence, Schopf told McKay it was far short of what they would need to convince scientists that there had been life on Mars at one time. Considerably more work would be needed.

Not only was the carbonate by itself insufficient, but they had to prove that the meteorite had not been contaminated as it sat in the ice in the Antarctic for 13,000 years. Furthermore, although the carbonate globules looked like those that are associated with life on Earth, carbonate can also be produced inorganically. They had to prove that it was produced organically.

McKay's team went back to work. What they needed now was a different way of testing the material, a different way of looking at it, and they found what they needed at Stanford University. Richard Zare of Stanford was an expert in laser analysis of carbonates and other minerals. McKay sent Zare and his team several rock chips for analysis. They bombarded them with laser beams and tested the resulting gas emissions. And when the results came back McKay was delighted. Deep within the globules were polycyclic aromatic hydrocarbons (PAHs). PAHs are made up of hydrogen and carbon; they are frequently found in fossil fuels and on barbeques (in overcooked meat), but of most importance, they are a by-product of living matter. When microbes die they leave PAHs.

The team was delighted but realized that PAHs also arise in inorganic processes; in other words, they aren't always associated with life. They have been found in other meteorites and are common on Earth, even in

Antarctica, so they could have been caused by contamination. Zare and his group therefore looked at them closer. They soon found that the PAHs in ALH84001 were slightly different from the ones they had commonly encountered in the past. They had a simpler structure than those formed inorganically on Earth. Zare was soon convinced that they were indeed the decay products of microbes.

But there was still the question of contamination. Zare looked at the distribution of PAHs in the meteorite and found that the highest concentration was near the center. If they were a contaminant from the ice fields, the highest concentration would have been near the surface.

Martian Worms?

The thrill of the discovery of PAHs hadn't worn off when McKay and several colleagues made another important breakthrough. McKay heard about a new and particularly powerful scanning electron microscope that had been installed in the engineering division at NASA. It could magnify up to 200,000 times. He requested and received permission to use it.

McKay and Gibson set up the samples from ALH84001 and adjusted the screen. When the image came into view they both froze in their seats. Strange, wormlike structures were clearly visible. They had a segmented structure and were tiny—about 1/100 the width of a human hair. Furthermore, there were egg-shaped structures in the same images. If they were, indeed, fossils of primitive life forms, you would expect the eggs to be nearby.

"I had difficulty sleeping that night," said Gibson. "It was the most exciting thing I've done in my 27 years as a scientist." They showed the images to the other members of the team and they all agreed that they looked like microfossils. Although they were much smaller than corresponding microfossils on Earth, there was no doubt about their similarity.

This is the part of the evidence that has garnered the most publicity, but McKay says that it isn't the strongest part of the evidence. He admits that tiny inorganic structures do form, but says that few are shaped like the objects they saw in ALH84001.

By now Thomas-Keprta's research was beginning to pay off. She had taken numerous TEM photographs of both the carbonate globules and the dark material that formed the rims, and she had identified some of the materials present. Tiny amounts of iron oxide and iron sulfide were de-

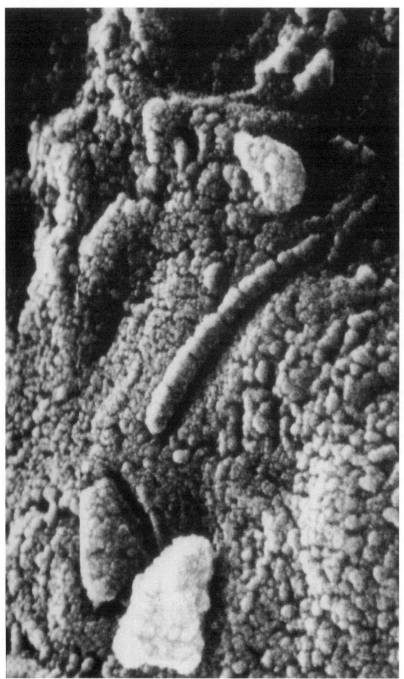

"Martian worm" found by McKay's team. (Courtesy NASA)

tected on the rims. Then, inside the globules she found greigite and pyrrhotite, both of which are common by-products of bacteria and other microbes on Earth. Magnetite was later identified; it is also frequently a by-product of life.

The team was now convinced that they had enough evidence to make their discovery public. They had structures that looked like microfossils that had been found on Earth, they had PAHs and other by-products of microbes, and they had the proper conditions and temperatures for life to form. In writing up their report they purposely kept it low-key, titling their article, "Search for Past Life on Mars: Possible Relic Biogenic Activity in Martian Meteorite ALH84001." The article was published in *Science* in August 1996.

Before the article was published, however, news leaked out and everyone was scrambling to find out some of the details. A news conference was therefore held on August 7, 1996. David Goldin, the head of NASA, began the conference with the statement, "NASA has made an important discovery." He referred to the evidence as "exciting, compelling, but not conclusive." He cautioned the audience that he wasn't referring to higher forms of life, but tiny, single-celled creatures that resemble bacteria on Earth.

McKay then took the podium. He discussed the evidence in detail, outlining the four lines of evidence that they had, and talking about the care they had taken in obtaining their data.

The Critics

The announcement of the possibility of primitive life on Mars was so dramatic that everyone expected severe criticism of the evidence, and indeed it came quickly. One of the chief concerns was contamination. The meteorite sat on the surface of Earth for 13,000 years and could have become contaminated. Although it was in Antarctica, which is relatively sterile, bacteria and other microbes do exist there.

Robert Gregory, a geologist at Southern Methodist University in Dallas, speculates that the water got into the meteorite when it sat on the ice in Antarctica, rather than when it was on Mars. The meteorite was dark and would therefore have absorbed considerable heat. According to Gregory this may have fractured the rock. Zare had pointed out that there were few or no PAHs on the outside, with most deep within the meteorite. Gregory believes that he can explain this. He said that water from melting snow could have crept through the fissures into the center of the rock, and become concentrated. He said that there were few on the outside because ultraviolet light from the sun would destroy them here. Others are now looking into the merits of Gregory's viewpoint.

Jim Papike, an expert on meteorites at the University of New Mexico (UNM), examined iron sulphide minerals in the meteorite. In particular he measured the ratio of two sulphur isotopes, and he found that they were not consistent with organic materials. Working with Chip Shearer, also of UNM, Papike tested thin slices of the meteorite. Shearer said they were disappointed, because they had hoped to find evidence of life. "[Our data] don't mean they are wrong," said Papike. "They just don't support [their conclusion]." It is possible that there are forms of life that don't leave the fingerprints he and Shearer were looking for, Papike admitted.

Ralph Harvey of Case Western Reserve University in Cleveland, who has combed Antarctica in search of meteorites ever since he was in graduate school, and Harry McSween of the University of Tennessee, working with John Bradley of MVA in Norcross, Georgia, cut a thin slice from one of the carbonate globules, then eroded it with a jet of argon ions until it was many times thinner than a human hair. Putting the slice under an electron microscope they found a crystal growth pattern known as an axial screw dislocation. The only place on Earth this type of structure is found is in volcanoes. They believe this is evidence that the carbonate formed at higher temperatures, perhaps as high as 1200°F. Harvey and McSween also measured the abundances of various elements in the carbonate regions. They found calcite and dolomite together, which to them suggested a temperature of at least 1200°. Furthermore, they believed the manganese–iron carbonate blend in the meteorite is only consistent with high temperatures.

On the basis of this, Harvey and McSween proposed a different route to the carbonates in the meteorite. They believe that the asteroid that struck Mars, throwing ALH84001 into space, created a hot fluid at a temperature of 1200° that was rich in carbon dioxide. According to them it was this fluid that deposited the carbonate in the meteorite as it was flung into space.

In another study Luann Becker and several colleagues at the University of California, San Diego, used a mass spectrometer to analyze the

PAHs from the Antarctic ice fields in the Allan Hills region. She found PAHs in these samples, and from other areas of Antarctica.

Another bone of contention is the size of the microfossils. Although they look like microfossils found on Earth they are a hundred times smaller than the smallest regular bacteria found on Earth. There are problems with this argument, however. Recently, microfossils on Earth of what are called nanobacteria have been found, and they are much smaller than regular bacteria. Little work has been done in this area so far.

The age of the carbonate globules is also controversial. Early estimates put it at 3.6 billion years old, and that has generally been accepted by the NASA scientists. But Meenakshi Wadhwa of the Field Museum in Chicago and Gunter Lugmair of the Scripps Institute of Oceanography believe it is much younger, on the order of 1.3 to 1.4 billion years. This, interestingly, is the age of several of the other Martian meteorites.

More Evidence

Despite the criticisms, McKay's group was given a boost on October 31, 1996, when a British team announced that they had identified organic compounds in another Martian meteorite. The team, which included Ian Wright and Colin Pillinger of the Open University and Monica Grady of the National History Museum, studied meteorite EETA19001, a basaltic rock that was formed on the surface of Mars only 180 million years ago. It was blasted into space a mere 500,000 years ago, and is therefore considerably younger than ALH84001. Of particular importance was the discovery that the ratio of carbon-13 to carbon-12 abundance in the organic matter was similar to that found in organic matter on Earth. The ratio of these two isotopes is different in living matter than it is in nonliving matter. This was not conclusive proof but it was very suggestive.

Amino acids were found in the meteorite, but a problem soon developed in relation to them. G. D. McDonald and J. L. Bada presented evidence that the amino acids are contaminants from Earth. They suggested that the acids got into the meteorite via water from the surrounding ice in Antarctica. Ian Wright and his colleagues pointed out that this was unlikely, however. They calculated that over a million cubic centimeters of water would have to flow through the rock to deposit the concentration of amino acids detected. If this had happened "it would be surprising if any of the rock survived," wrote Wright.

Countering the Critics

McKay admitted at the press conference in August 1996 that their evidence was tentative, with each piece of evidence by itself relatively weak. Taken together, however, he believes it is exceedingly strong.

Gibson also feels that their evidence will prevail. "We feel very strongly that everything we have said is valid and has recently been strengthened," he said. Considerable new evidence was presented at the twenty-eighth Lunar and Planetary Science conference at JSC and the Lunar and Planetary Institute in Houston in March 1997. Papers were presented from both sides of the argument, but most (about 70 percent) were in favor of McKay's view. Joe Kirschvick of Caltech, the discoverer of magnetite in living creatures, showed that the magnetite found in ALH84001 has a different crystalline structure from the magnetite found in nonbiological systems. He also showed that they had to have formed at temperatures below 150°C (302°F).

Allan Treiman of the Lunar and Planetary Institute looked at Harvey and McSween's arguments that the carbonates formed at high temperatures and he believes they are flawed. He pointed out that they are assuming the system is in chemical equilibrium, and he says it is not.

Several researchers, including McKay, Thomas-Keprta, and Gibson, have looked for possible analogs of the Martian microfossils on Earth, and they have found forms that they believe are nanobacteria. They were isolated from lava rocks beneath the Columbia River plateau in eastern Washington and are comparable in size and shape to those found in ALH84001. They also searched in thermal springs, in particular in a spring in Italy, and discovered clusters of spheroidal objects that looked like bacteria. They were also similar in size to the Martian microfossils. Another group found similar fossils from 3.5-billion-year-old rocks from South Africa. Hojatollah Vali of McGill University and several colleagues found bacterial systems on Earth that produced the same minerals (magnetite and iron-rich carbonates) as in the carbonate globules in ALH84001.

McKay pointed out that bacteria on Earth usually produce thin films of organic polymers called biofilms. He and his team used an electron microscope to search for these films. They reported that they observed films near the microfossils in ALH84001 that appeared similar to the terrestrial biofilms.

In a more philosophical vein, Treiman cautioned scientists about the dangers of trying to extrapolate from Earth to Mars. "It is dangerous to analogize and extrapolate from Earth to possible life forms in the poorly known environments of Mars," he said. "Our knowledge of Mars is basic; our knowledge of Earth's life is growing; nearly all past extrapolations to extraterrestrial life have been wrong."

Despite the evidence, there is, understandably, still some skepticism. Is there any way, you might ask, with the present evidence, that we can prove once and for all that we are, indeed, observing life from Mars? One way is to cut into the microfossils and examine them under the electron microscope for cell walls and other cellular structures. If they are, indeed, fossils of microbes, they should show such structures. This work is still in progress.

Even if McKay and his team are eventually proved wrong they have made an important contribution. As a result of their work there is a renewed interest in the quest for life on Mars. NASA has, in fact, moved several of its Martian missions forward because of it.

chapter 7

Mars: The Future

Anticipation was high. *Mars Observer* had been launched successfully several months earlier and was now approaching Mars. It would go into orbit in only a few days. *Observer's* cameras were focused on the red planet, and scientists at NASA were glued to screens around the room, watching as the image of Mars grew larger and larger. Then, as they prepared for the landing, the transmitter died and *Observer* went out of control. An aneurysm in a fuel line and a billion-dollar gamble was lost: *Observer* went spiraling off into space. The mission was a failure.

The engineers and scientists at NASA were devastated, and in Washington government officials were shocked and dismayed. Talk of cutbacks ensued. Something had to be done. The missions were too expensive and too much of a risk. Still, it was twenty years since we had visited Mars and another visit was overdue.

A new approach was needed. In a bold move, Daniel Goldin was brought in from industry to head up NASA. Goldin had gone through a system where budgets were tight and little money was wasted. He promised that he would take the same approach at NASA.

Within months there was a new program. The billion-dollar Cadillac spacecrafts would be replaced with smaller, lighter, and cheaper crafts, each costing only $150 million. NASA engineers were skeptical. How could you get to Mars on $150 million when it had cost a billion dollars previously? It seemed impossible. Then the plans were laid out. Most of the metal in the spacecrafts would be replaced with lighter, stronger, synthetic materials, and the fuel load would be severely decreased. Considerable fuel had been taken all the way to Mars in previous flights for use in slowing the craft as it entered the Martian atmosphere. Most of this fuel would be done away with. New techniques would be used for braking. And finally, without sacrificing sophistication and accuracy, the instrumentation would be made smaller and lighter. With these adjustments the new crafts would weigh only half that of the earlier spacecrafts.

Despite the cutbacks little would be lost. The cameras and other equipment aboard the new spacecrafts would be state of the art, and vastly superior to those aboard *Viking*. Furthermore, if one was lost, it would not be a financial disaster for NASA.

Ten such missions were planned—two every two years when Mars and Earth were in opposition.

Global Surveyor

The first of the new crafts, *Global Surveyor*, was launched in November 1996. It had a weight of 2337 pounds, compared with the 5672 pounds of the earlier *Mars Observer*. A new technique called aerobraking was used for slowing the craft down when it got to Mars. It was first tried at the end of the Magellan mission to Venus and was successful, but this was the first time it was an integral part of the mission.

Aerobraking is not new; it has been around for years. But it was something that was looked upon as a nuisance, something to be avoided. It shortened the life of Earth satellites. When engineers put a satellite into orbit above the Earth they try to make the orbit as circular as possible. Why? Because elliptical orbits create problems: at some point in the orbit the spacecraft passes well into the Earth's atmosphere and air friction slows it down. This causes it to go into a new orbit closer to the Earth; and each time it passes through our atmosphere its orbit is decreased until finally the spacecraft is overcome by air friction and crashes to Earth.

Highly elliptical orbits are disastrous for Earth satellites, but for *Global Surveyor* a highly elliptical orbit was desirable. When the ship arrived at Mars on September 11, 1997 its velocity relative to the Martian surface was high—exceedingly high—and it had to be slowed down if it was to go into orbit. It was therefore put in an elliptical orbit so it could use air friction to slow it down. The objective was, of course, to change its orbit gradually into a circular one as close to the planet's surface as possible. This approx-

Global Surveyor orbiter. (Courtesy NASA)

imately circular orbit will be achieved in January, 1999; at this point *Surveyor* will be about 235 miles above the surface.

Surveyor has a sophisticated array of equipment aboard. Its cameras give images that are ten times sharper than those of *Viking*; they can resolve objects on the surface as small as five feet across. Thousands of pictures will be returned to Earth. Furthermore, it has several instruments in addition to cameras. A thermal emission spectrometer will analyze infrared radiation from the surface, and a laser altimeter will send a laser beam down to the surface to measure the heights of mountains and the depths of valleys on Mars, giving us a good topographical view of the planet.

In addition, a magnetometer will search for a remnant magnetic field. Mars is considerably smaller than Earth and has cooled more, and therefore has a much thicker crust. Little is known about what is below it, but we know that at one time it had a relatively large molten core; the large volcanoes on the surface attest to this. Estimates suggest that some of the volcanoes were active as recently as 200 million years ago, and some may still be active today. And if Mars had a molten core at one time, it should still have a small magnetic field. According to *Viking* results, however, it would have to be very small. Furthermore, if there was molten lava close to the surface in places there may be liquid water—underground thermal springs—in these regions. *Surveyor* will be on the lookout for evidence of them.

Surveyor is expected to be operational for two years. During that time we will no doubt learn a tremendous amount about Mars.

Pathfinder

Strangely, even before *Surveyor* became operational another spacecraft, *Pathfinder*, was already on the surface. *Pathfinder* wasn't launched until December 1996, but it reached Mars a full two months before *Surveyor*. It was lighter and therefore obtained a greater thrust from its booster rocket, giving it a greater velocity through space. *Pathfinder* reached the surface of Mars on July 4, 1997.

Unlike *Mars Observer* and *Viking*, *Pathfinder* was not just a lander; it had a full array of instruments and cameras for studying the surface and weather, but it also had a very special feature—a small six-wheeled rover that could move about in the area around the lander, studying the rocks and soil.

Mock-up of Pathfinder on surface of Mars. (Courtesy NASA)

Sojourner being readied for flight. (Courtesy NASA)

Before *Pathfinder* was launched, NASA had had a competition in which students from across the nation were asked to submit a name for the small rover. (They had to submit an essay justifying their selection.) The winner was *Sojourner*. A sojourner is a traveler who stops for a time before moving on, and this is exactly what the rover was designed to do. The name was also in honor of the African-American reformer Sojourner Truth, who lived about the time of the American Civil War and traveled from town to town fighting for the rights of all people to be free.

One of the most important decisions for the *Pathfinder* engineers was the selection of a landing site, and there was considerable discussion about it. A region that was relatively smooth—free of large rocks and so on that could disable the craft—was needed, but it also had to be a region where there had been water in the past, because that would be the most likely place to find indications of life on the planet. Neither *Pathfinder* lander nor *Sojourner* would be able to see or identify life (microfossils or microbes); neither was equipped with the sophisticated biology experiments that *Viking* had aboard. Nevertheless, they would pave the way for later flights that will be capable of detecting life.

The site also had to be in the lowlands so the atmosphere would be thick enough to support the parachute as the craft landed, and there had to be considerable sunshine to power the solar cells the craft would use later. The team finally settled on a large flood plane called Ares Valles, or Valley of Mars.

Unlike *Viking*, which went into orbit when it reached Mars, *Pathfinder* went directly to the surface. After a brief blast of its retrorockets to slow it down a parachute was deployed to take it to the surface, then seconds before it struck, a large cocoon of airbags inflated, which cushioned its landing.

It bounced across the surface using up its remaining kinetic energy. Engineers were worried that it might come to rest upside down, but as it turned out, it was upright, with the base down and the antenna up. Furthermore, radio contact was not lost. Once it was at rest the airbags were deflated, and within a short time the three petals that make up its shield were released and flattened on the ground. *Sojourner* was attached to one of them—folded like an accordian to save space. It stood up to its full height as the lander surveyed the region around it with its cameras. Engineers back at Jet Propulsion Laboratory (JPL) were pleased; the view was astounding. They looked over the area around the lander to see where the best route for *Sojourner* was.

Although *Sojourner* gave us a tremendous amount of information, the lander was not relegated to a secondary role. It had a large array of instruments. The three triangular-shaped petals were covered with solar cells, and one of them housed a weather station that was equipped with wind socks placed at various heights to monitor surface winds; it also had several temperature gauges. Between the three petals was the base. It was about nine feet across and had a camera on its mast that stood approximately five feet above the ground. This is considerably higher than *Viking*'s camera, and gave us a better view of the region around the lander. The camera gave stereo images and was equipped with a set of filters for examining the landscape at various wavelengths.

Sojourner was equipped with three cameras. The entire top of the small vehicle was a solar array, but it also had batteries for use at night. Its six steel wheels were designed to move independently over a rocky uneven surface and gave the vehicle considerable stability. It was small—only two feet long by one and a half feet wide and its wheels were a mere five inches in diameter. Nevertheless it was well equipped to provide a lot of information.

Sojourner moved out across the surface at a relatively slow pace—less than two miles per hour. Each night it parked near a rock and studied its properties using an alpha X-ray spectrometer. By mid-September it had

The view from Pathfinder. The resolution is better than Viking photos. (Courtesy NASA)

Sojourner on the surface of Mars checking a rock. (Courtesy NASA)

analyzed eight rocks. They were given names such as Barnacle Bill, Shark, and Yogi. One of the major surprises was the high silica content of several of the rocks. In this respect they were quite similar to rocks on Earth. Their high silica content indicated considerable crustal activity which in turn meant that the Martian surface had been heated and recycled several times. Another group of rocks had high sulfur content. Finally, several rocks and the soil were shown to have a chemical content close to that of the Martian meteorites, giving further evidence that ALH84001 is, indeed, from Mars.

A second surprise was the rapid temperature changes that can occur just above the surface. Temperatures can plunge 40° F in a matter of minutes. *Sojourner* also found considerable evidence that a large flood had occurred in the area; the boulders around it were all facing in the same direction, and there was evidence of layered sediment. Of particular interest, it also uncovered evidence that Mars was once much warmer and wetter, strongly suggesting that it may have had life on its surface in a earlier era.

Engineers and scientists at JPL were pleased with the results. *Sojourner* was designed to operate for a week and the lander for 30 days, but they returned images for 84 days. In all, the lander returned 16,000 images and *Sojourner*, 550.

The Next Step

As *Pathfinder* and *Global Surveyor* become history, others will be coming to replace them. In 1998 two more spacecrafts will head for Mars: *Mars 98 Orbiter* and *Mars 98 Lander*. *Orbiter 98* will continue to photograph the surface and measure the water and dust content of the atmosphere. The lander will land near the south ice cap and dig a trench, looking through the layers of ice—back in time. Ice is like the layers of a tree, and much can be learned from it. The south polar cap is one of the most promising places for organic remains of Martian life.

Japan is expected to launch a rocket to Mars at about this time. Called *Planet B*, it will measure the properties of the upper atmosphere.

Then in 2001 another two crafts will go, again an orbiter and a lander. The orbiter will carry instruments to look for ice or possibly liquid water beneath the surface. So far all landings have been on the smooth regions of

Mars. But the highlands are much older and are of considerable interest. A landing in this region is planned for 2001.

The much-awaited mission is the one that will come in either 2003 or 2005. It is a sample-return mission. In this case the lander will scoop up some soil, then it will be boosted back into space and eventually come back to Earth. The big question, of course, is where to take the soil samples. Chris McKay of NASA/Ames Research Center in California (no relation to JSC geologist David McKay) has been looking into this. Over the past few years he has searched the polar regions of the Earth for Marslike conditions, examining the life beneath the surface at each site. He believes the most promising sites for life forms are ancient dried lakebeds, former hydrothermal springs, and the permafrost near the south pole. The sites of former hydrothermal springs would likely be difficult to find, but it is possible that some still exist. They are important because hot mineral water tends to fossilize living organisms.

The permafrost near the south pole is an excellent target in light of a number of recent discoveries. Russian scientists have found microorganisms in the Siberian permafrost that are still alive after being frozen at 15°F for 3.5 million years. On Mars there is the problem of radiation, so it's unlikely we'll find living organisms, but dead ones would still be a significant discovery.

Man on Mars

Our ultimate aim is, of course, to get a man to Mars, and we may have to wait for this mission to resolve the question of whether there is life on Mars. This mission will not come for a few years, but there's little doubt that it will eventually come. The problems involved in putting a man on Mars are much greater than putting a probe on the planet. The support payload for man is much greater, and the spaceship would therefore be much heavier. We've already gone to the moon, but a trip to Mars is considerably longer—of the order of six to eight months to get there and the same amount of time to get back—and would be much more difficult. One of the problems would be zero gravity. Degeneration of muscle and demineralization of bone occurs when there is no gravitational field. Muscle degeneration can be overcome to some degree by extensive exercising, but demineralization of bone cannot. Once the astronaut is back in a gravitational field, however, bones eventually build back. One way to overcome this is to create an artificial gravitational field. This could be done by rotating the spaceship, but effective ways of doing this are still far in the future.

The main reason for sending a man to Mars rather than a probe is that a man could do a much more thorough job of exploring. Mars is so much smaller than Earth it might seem that we would have little trouble exploring it, but in reality there's as much land area on Mars as there is on Earth. The oceans take up much of the area on Earth.

The question of life is paramount in relation to Mars, and early explorers would spend much of their time searching for evidence of it. They would also likely spend considerable time looking for microfossils, but it is possible there are also fossils of higher forms of life. Dinosaurs roamed the Earth millions of years ago and, with the exception of a few fossils, we see little trace of them now. There may have been a similar situation on Mars. Higher forms of life could have roamed over its surface millions or perhaps billions of years ago when it had a thicker atmosphere and flowing water. So fossils of higher forms of life may exist.

But if we are to explore much of the planet we would have to be mobile, and that would mean we would need a vehicle to take us from place to place. The astronauts of the *Apollo 15* mission to the moon used a vehicle, and it proved to be helpful. What type of vehicle would be best? There are two alternatives: an electrical vehicle, powered by batteries or solar cells, or a combustion vehicle, similar to cars on Earth. At the present time combustion vehicles are still far more efficient than electrical ones, but that may eventually change. The major problem with combustion vehicles is the propellant. What would we use for fuel? We could bring the fuel needed for the first few trips to Mars, but eventually another source would have to be found. Fortunately, there is a source on Mars. The Martian atmosphere is 95 percent carbon dioxide, and carbon dioxide combined with various other gases is an excellent fuel. A mixture of carbon dioxide and hydrogen, for example, would be extremely efficient. There is no hydrogen on Mars so it would have to be brought from Earth, at least initially. This is not a problem, but storing it for long periods of time is.

Another possibility is a mixture of oxygen and methane. Oxygen could be obtained from the carbon dioxide in the atmosphere, and methane is relatively easy to produce. Carbon dioxide and hydrogen combine to form methane and water, and water can easily be split into hydrogen and oxygen. According to the experts, a mixture of oxygen and methane may be our best bet.

With the problem of the fuel out of the way, our next problem would be navigation. Mars has little or no magnetic field, so a compass wouldn't be of much help. Furthermore, Mars is much smaller than Earth and line-of-sight radio communication with a base would be impossible over about 25 miles. Fortunately we can do as we do on Earth: Mars has an ionosphere and we could use it to reflect radio signals. Also, large numbers of satellites will no doubt eventually orbit the planet, and they could be used for guidance.

Building a Base

A permanent base on Mars will be a necessity if we are to explore it properly and consider it for eventual colonization. Of all the planets in the solar system, Mars is the only one we are likely to colonize. Venus is a possibility, but it is far too hot and hostile, and it doesn't have the natural resources that Mars does. Our moon is also a possibility, but it has few natural resources compared to Mars.

There are two main problems in setting up a base on Mars: we can't breathe the air, and we would need protection from the ultraviolet radiation that penetrates to the surface. Neither of these are serious. Large plastic domes could be brought from Earth and a simulated Earth atmosphere could be produced. Furthermore, the domes could be made to protect people from ultraviolet radiation.

As the base grows it would have to become self-sufficient; everything couldn't be brought from Earth. We would have to rely more and more on the material resources on Mars, and indeed there are a lot of them. The base could eventually become self-sufficient, or at least close to it.

One of the first things we would need on the base is water. As we saw earlier, there is considerable water on Mars. Unfortunately, none of it is sitting on the surface. The two main sources are the permafrost layer beneath the surface, and the polar ice caps. We are not likely to build our first base on or near a polar cap, so the best region is where we are certain there is considerable subsurface ice, namely, a region where there was an ocean at one time. As I mentioned earlier, it is also possible that water exists in liquid form beneath the surface; in other words, there may be geothermally heated pools of water. These springs are one of the things early explorers will search for. And it is quite possible they exist; after all, we know there is considerable water in the frozen state beneath the

Transportation depot in Mars orbit. (Courtesy NASA)

surface, and that volcanoes were active in the past, so there may be lava close to the surface that would melt the ice.

If we don't find underground pools, another source of water is the permafrost, but it won't be easy to mine, and it may be years before we develop techniques to extract it efficiently. In the meantime, however, there is another source. The Martian soil appears to be as dry as anything we see in the Sahara on Earth, but according to *Viking* results it is about 1 percent water by weight. This may not seem like much, but there's a lot of soil and it's easy to get at. Furthermore, it is believed that in places the water content of the soil may be 3 percent or more. And surprisingly the water is easy to get out of the soil; all you have to do is heat it.

Once we have a base with water it's natural to think of cultivated crops. According to *Viking* the soil has a good supply of nutrients, and we should be able to grow things in it. Mars is farther away from the sun than Earth, so sunlight is not as strong (it is 43 percent that of Earth), but this is sufficient. Furthermore, there are few or no clouds on Mars. Large greenhouses could be set up that would help make the base self-sufficient. Almost anything could be grown there that can be grown on Earth.

Another important requirement of a base would be power. Would it, in fact, be possible to generate sufficient power on Mars to sustain a base? There are several possible sources of power. Initially, solar power would likely be the main source, but there is considerable wind on Mars and wind power could be developed easily. The atmosphere is thin, but wind velocities, particularly at high altitudes, are substantial and considerable energy could be generated using windmills.

Geothermal power is another possibility if underground thermal pools are found. Geothermal power is a major source of power in certain areas on Earth (e.g., Iceland), and if available on Mars, it would give us a tremendous advantage. Finally, nuclear power could also eventually be developed.

With a good source of power it would then be possible to manufacture things, particularly plastics. Plastic is a material that is used extensively on Earth, and it would be needed on Mars. The large domes, for example, would have to be made of some form of plastic. Again, with a little ingenuity we would have no problem manufacturing plastics. All the basic ingredients for it are there. Ethylene is the main component, and it can be obtained from carbon dioxide and water (or hydrogen). A plastic manufacturing plant is not beyond our means. Within a few years of establishing a base one could easily be built.

And if we could make plastics we could also manufacture metals. One of the main components of the Martian soil is iron; it's rusty iron, in fact, that gives Mars its color. With iron it would be possible to make steel. Aluminum is also present in the soil.

Terraforming

If bases were set up on Mars and it was eventually colonized the next step would be terraforming, in other words, changing its temperature and atmosphere to make it more like Earth. This would not be a short-term project. It would take a long time to convert Mars's atmosphere over to the one we have on Earth, but it's not impossible.

The atmosphere of Mars is much thinner than the Earth's, and temperatures are well below those on Earth. We would also want running water on the surface, but that would come as the atmosphere was thickened. Our first job would be to increase the atmospheric pressure. This could be done by placing a large mirror in orbit around the planet. Sunlight could be directed at the polar caps; they are composed of both frozen carbon dioxide and frozen water. As the carbon dioxide sublimates into the atmosphere its pressure would increase. The water ice would evaporate as water vapor and help increase the pressure. Eventually the pressure would be high enough so that when the ice melted, water would flow onto the surface.

As the atmospheric pressure built up, the greenhouse effect would help heat the surface, and temperatures would soon be close to those on Earth. Furthermore, as the water flowed across the Martian soil, oxygen would be given off. This oxygen would go into the atmosphere, eventually making it possible for humans to breathe the air. In addition, the oxygen would create an ozone layer in the atmosphere that would protect the surface from the deadly ultraviolet radiation.

How long would such a process take? Estimates vary from a few hundred to perhaps 600 or 700 years. This may seem like a long time, but what is important is that it is possible.

chapter 8

Other Life
in the Solar System

To many, Mars is the most likely object in the solar system to harbor life, even though we now know that it can only be an elementary form. But Mars is not the only object that may contain life. One of the moons of Jupiter, Europa, has also attracted a lot of attention lately. A number of scientists have, in fact, speculated that it is more likely to harbor life than Mars. Despite its location in deep space, far beyond the life zone of the sun, Europa has many properties that make it a desirable candidate. It has more water than Mars, and most of this water *appears to be in a liquid form*. Furthermore, the water is in a protected environment, with a layer of ice over it.

And beyond Europa is another candidate—Titan, the largest moon of Saturn. Titan is considerably larger than Europa, and it has something that Europa doesn't have: a dense atmosphere with considerable nitrogen in it. (Nitrogen is the main component of Earth's atmosphere.) Unlike Earth, Titan is unbelievably cold, but it may be the only place in the solar system besides Earth that is partially covered with liquid that is not under ice.

Scientists are convinced that both Europa and Titan are excellent candidates for life, but there are other objects of interest in the solar system. The probability that they harbor life is not particularly high, but they are worth looking into. Ganymede, the largest moon in the solar system (and like Europa, a moon of Jupiter) also appears to have water below its surface, as does Callisto, another of Jupiter's moons. Neither is considered to be an excellent candidate, but with water, they are both of interest.

Two planets (other than Mars) are also of interest, namely, Jupiter and Venus. Scientists have speculated for years that either of them could contain elementary forms of life. At first glance neither looks like a good prospect; Jupiter is mostly gaseous and Venus has a surface that would fry any form of life. But on both planets there are regions in the atmosphere that have moderate, Earthlike temperatures. In addition, both have many of the gases that were in the primitive atmosphere of Earth.

Finally there are the asteroids, meteorites, and comets. We have, indeed, already found amino acids and some of the building blocks of DNA in meteorites and it is expected that comets may contain the same molecules.

Europa

Europa lies far out in the solar system, so far that it is barely warmed by the sun's rays. From its surface the sun would be a small, glowing white ball in the sky. The temperature on the surface of Europa is $-270°F$— unimaginable to us here on Earth. The lowest temperature ever recorded on Earth is a mere $-128°F$. It may surprise you that Europa, with temperatures this low, could be a good candidate, but as we will see it's not what's above its surface but what's beneath it that counts.

In 1979 the second of the two *Voyager* spacecrafts swung past Europa, and the first images of the moon were transmitted back to Earth. They showed a strange, smooth object crisscrossed with long dark lines. Nothing like it had ever been seen before. But just inside its orbit was the even more intriguing moon Io, with its colorful, pizzalike surface, and over the next few months it overshadowed Europa. Europa was lost in its glow. But finally attention was drawn back to Europa. Its surface was a mystery; it looked like ice, but how could we be sure? And what was below the ice? Speculation abounded for years. Then in 1996 came the even more startling images from the satellite *Galileo*. Surface features stood out clearly. There were two types of terrain: mottled regions of grey and brown, and large bright flat regions. The dark regions are believed to be rolling hills of ice, but they are not high—at most a quarter of a mile above the smoother regions of the surface. Europa is, in fact, the smoothest object in the solar system.

The second type of terrain is flat and reminiscent of the large sections of frozen ice in the Arctic. It is different, however, in that it is crisscrossed

Europa from 150,000 miles above the surface. (Courtesy NASA)

by long dark lines—some as long as 1000 miles, and some as thick as 30 miles. In the early 1980s, shortly after the *Voyager* visit, NASA scientist Patrick Cassen suggested that they were cracks in a surface of ice, and this is now the accepted interpretation. But because of their thickness, something else has to be going on.

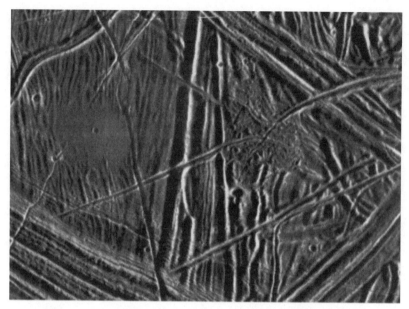

Close-up of Europa showing structure of surface. (Courtesy NASA)

Overall the surface is young, but there is considerable controversy as to how young it is. Before the latest images it was believed to be about 30 million years old. Now we are less certain. Clark Chapman, a planetary scientist with the Southwest Research Institute in Boulder, Colorado, believes it may be only a million years old. Others, however, do not agree. Michael Carr, a geologist with the U.S. Geological Survey, believes it is far older.

Anyway, scientists were fascinated as each new group of images came in (*Galileo* made several passes of the moon). There was a tremendous variety in the features and markings on the surface; long strips that looked like superhighways, regions that looked like icebergs locked in a frozen sea. What did it all mean? Where did they come from? To answer these questions it is best to look back at Europa's history: how it developed and evolved. And since its history is tied to the history of the sun and other planets of the solar system, we'll begin with them.

Through intense study and careful speculation we now believe we have a fairly accurate picture of how the solar system came into being. We know the Earth is approximately 4.6 billion years old (time since its surface

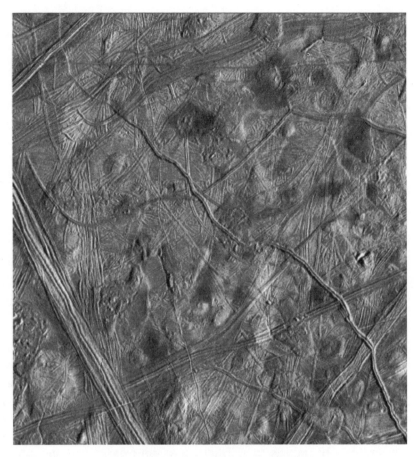

Cracks and other structures on the surface of Europa.

solidified) and that the solar system began to take shape about 5 billion years ago. The story begins with a huge swirling gas cloud. This cloud was much larger than our solar system, reaching almost a quarter of the way to the nearest stars (about a light-year). It was made up mostly of hydrogen and helium, but contained a small fraction—less than 1 percent—of heavier elements. We refer to it as the solar nebula.

This solar nebula was irregular in shape at first, the remnant of a star that exploded violently in the distant past. Gravity pulled it inward, and it became more spherical, and as it continued to fall its spin increased and it began to flatten. Eventually it became a huge spinning disk with a dense, bulging center. Over millions of years, the sun and planets would evolve out of this cloud of gas and dust.

Looking at the solar system today, we see what may appear to be a strange array of planets: four tiny closely spaced planets with solid surfaces near the sun, and four gas giants beyond them (we are ignoring Pluto for now). How did this distribution come about? The answer lies in the continuing evolution of the solar system. As matter fell inward as a result of self-gravity, the core of the nebula grew increasingly dense. Deep within it pressures mounted and the gas got hotter and hotter. This is the region that would eventually become the sun, but at this stage there were no nuclear reactions in the core; it was heated by compression in the way a ball of putty is heated when you squeeze it. We refer to it as the protosun.

The heat could not be contained and gradually began to move outward, warming the huge disk-shaped envelope that surrounded the protosun. This is the region that would eventually become the planets. The gas close to the protosun heated until it was too hot for ice to exist. Further out, however, it was cooler. In fact, there was a gradual drop-off in temperature as you moved outward through the nebula.

As the system continued to evolve, grains—like the grains of sand on a beach, but much smaller—began to condense out of the cloud. Close to the sun, where the temperature was high, grains of iron, nickel, and silicate condensed; the silicate would eventually produce rock. This was the region the Earth was in.

As you moved outward, however, the composition of the grains changed. More and more of them became ice particles, ices of various gases, and yes—water ice. At a distance four times as far from the sun as Earth the grains were mostly water ice. It was these grains that would eventually collide and form the building blocks of the planets, objects we call planetesimals.

The planetesimals of the inner solar system were composed of rock and iron. They were similar to the rocks in space—the asteroids—we see today. The planetesimals of the outer solar system, on the other hand, were composed mostly of ice, similar to comets. But there were some heavy elements present, and they quickly fell to the center of the newly forming planets (protoplanets) in this region. Around these nuclei of rock and other heavy elements, ice quickly built up until the outer protoplanets were much larger and more massive than the inner ones. The inner protoplanets were in a region where there was no ice.

At this stage everything was still immersed in a dense cloud of gas, made up mostly of hydrogen and helium. And as the icy spheres in the outer regions built up, their gravitational fields strengthened and they began to attract gas from the nebula. Huge envelopes of hydrogen and helium were attracted to them, while only small amounts of gas were attracted to the tiny inner protoplanets.

The sun was now a huge glowing red ball—a fiery ember considerably larger and brighter than it is today. The gases within it were still being heated by compression. But gravity continued to pull it inward, and as it contracted it grew more dense, and its core grew hotter. Finally the temperature of the core reached 15 million degrees and nuclear reactions were triggered. Our sun was now a star.

But the reactions that made it a star triggered a shock wave that moved out from the sun, sweeping past the newly forming planets and cleaning out the gas from around them. All that remained was an array of four closely spaced rocky spheres, and four gas giants farther out. The shock wave was not strong enough to blow the large dense atmospheres from the giants; they were too far from the sun.

Europa was one of 16 moons whirling around the largest of these giants, Jupiter. It was far from the sun and received little heat from it. But Jupiter's system of moons was a miniature version of the solar system. Jupiter didn't have enough mass to become a star, but like the sun it condensed and was heated by compression. Farther out Saturn also began to glow dimly, but it was less massive than Jupiter, and therefore considerably dimmer. Still, for a short time there were two glowing embers in the outer solar system.

Jupiter heated the region around it as it glowed dimly but the only objects to enjoy this brief balmy period were its inner moons, Io and Europa. Both were now emerging out of the solar nebula, and both had condensed out in a blizzard of ice and were composed largely of water ice. Like Jupiter, however, they also had a rocky core.

Water ice, strangely, is a relatively common material in the outer solar system, more common even than rock. The inner cores of both Jupiter and Saturn are composed of rock, but around this rock is a thick layer of ice. Furthermore, a large fraction of both Uranus and Neptune appear to be composed of water ice.

Of Io and Europa, however, only Europa was ideally situated to take advantage of the radiant heat from Jupiter. Io was too hot, and it eventually lost its water. Temperatures on Europa, on the other hand, were moderate, and soon the surface was covered with a gigantic ocean—an ocean of water. Above the ocean was a thick atmosphere, rich in carbon dioxide, nitrogen, water vapor, ammonia, methane, and hydrogen. It was a brooding immense ocean swept by wind, its water whipped into towering

Artist's conception of early Europa.

waves. Then eruptions on the ocean floor began, and lava began to seep out, forcing its way upward through the cracks. Soon the floor was covered with active volcanoes spewing gas and lava into the ocean.

The turbulence above the surface continued, with rain pouring down into the ocean, carrying with it atoms and molecules from the atmosphere. Lightning was also prevalent. Looking at this we see it is similar to the conditions on the early Earth. Europa had a reducing atmosphere, a huge ocean, and ample sources of energy—lightning, ultraviolet light from the sun, volcanoes—everything that was needed for life.

With everything it needed in hand there is little doubt that complex molecules formed. But did the components of DNA, amino acids, and so on, form? We have no way of knowing for sure, but it appears that, if the process got started, it didn't get very far. Jupiter would never become a star; it was living on borrowed time—its heat was being generated by compression and would therefore be short-lived. The first life on Earth (fossils) appeared about a billion years after the Earth formed, and the tropical period on Europa lasted for only a few million years. But as Jupiter continued to cool there was something that delayed the inevitable, something that extended the period of warmth and moderation. The dense clouds above the surface trapped the radiation in the same way Venus's clouds now do. A greenhouse effect was set up, and it kept the surface warm for millions of years longer.

Still, Europa's fate was sealed. As Jupiter's long contraction came to an end, the radiant heat from its surface ceased. Today it still radiates two and a half times the amount of energy it receives from the sun, but this is of little help now and would have been of little help then. The oceans of Europa began to freeze; a wave of freezing swept across the gigantic ocean. The ice broke up, then reformed, but eventually the entire ocean was frozen solid, and as Jupiter cooled further, the ice thickened.

Did the water continue to freeze until it was solid throughout? It now appears as if it didn't, and to see the reason we merely have to look to Io. The most awe-inspiring discovery of *Voyager* was the surface of Io; on it were gigantic volcanoes that were spewing sulfur hundreds of miles above the surface. But as we saw earlier, Io has no water; because it was so close to Jupiter all its water evaporated. Still, it has a region of "liquid" sulfur beneath the surface that feeds the volcanoes. This region is created by tidal forces from Jupiter. You are no doubt familiar with tidal forces in relation to our moon. As our moon sweeps over the oceans it pulls them outward (even land masses are pulled toward it to a small degree). And if you went to the moon and examined its surface, you would find that it is bulging in the direction of Earth.

Jupiter has such a strong gravitational field, and Io is so close to it, that it pulls its surface 300 feet out from its usual position. This large bulge travels around Io as it rotates, resulting in a pulsing—a heaving of its surface in and out, back and forth. In this way Jupiter delivers an enormous amount of energy to Io, energy that creates heat and produces volcanoes.

The same pulsing action takes place on Europa. Europa is farther from Jupiter than Io so the effect is not as great. Still, it is large; the surface of Europa is pulled 30 feet in the direction of Jupiter, and the same pulsing that occurs on Io occurs on Europa. This action delivers a tremendous amount of energy to Europa. Was it enough to keep the ocean from freezing solid? From all indications it was. The surface continued to freeze until it was a few miles thick, then because of internal heating the freezing stopped. Furthermore, some of the energy was delivered to the region beneath the ocean, and the volcanoes that had formed remained active.

In the meantime, above the surface Europa's atmosphere grew thinner. The gravitational pull of Europa was insufficient to hold it; over thousands of years it leaked off to space. Today we find only a tiny remnant of oxygen left. But this didn't matter. The first vestiges of life were now protected far below the surface—beneath the ice in the water below.

We know that life began in the oceans of Earth, and here was an ocean that covered the entire moon, an ocean that was far deeper than our oceans—perhaps as deep as 60 miles. And within it were the first primitive molecules of life, formed during the brief balmy period. It was a dilute primordial soup. Furthermore, the 4 billion years that would come after this had barely begun.

Europa had everything that was needed: complex hydrocarbons, moderate temperatures, liquid water, nutrients, protection from ultraviolet radiation, and above all, a large amount of time. What would the outcome be? The chances appear to be high that some form of life would arise. We can't, of course, be sure, but the odds are on our side. If it did develop, however, it is likely to have been a primitive form. But there is always the chance that larger forms developed—fish and other aquatic life.

In connection with these larger forms, interesting objects called "black smokers" have been discovered on the floor of the Pacific Ocean, a mile and a half down in the darkest depths of the ocean—far beyond the point where sunlight reaches. A black smoker looks like the base of a burnt-out tree. They are called "smokers" because oceanographers have observed dense clouds of microbes rising from them. But crabs, small fish, and tube worms have also been observed in them, obviously sustained by heat and chemicals from the ocean floor. The conditions here may be quite similar to those on the floor of Europa's oceans, and many astronomers believe that similar objects could exist there. The life associated with them would likely be different, but the discovery of any form of life would be a fundamental discovery.

Do we have any other evidence for the picture of Europa that I have presented? And, indeed, if it is true, how thick is the ice? How far would we have to drill through to get to the water? Estimates vary, but most astronomers believe it is no greater than six miles thick, and recent images from *Galileo* indicate it may be as thin as a mile in places.

What about further evidence that the surface layer is, indeed, ice? We don't have to go far for this. The tidal forces that sweep over the surface not only keep the water in a liquid form but also affect the ice. What would happen to a huge sheet of ice if it were pulled outward by 30 feet? It would crack. Small cracks would appear that would quickly sweep across the ice at the speed of sound. Some of them would continue on for thousands of miles. Over time large numbers of cracks would appear.

The cracks would have another important impact on the surface. Beneath the ice is water, and gas from the volcanoes may also still exist there. They would exert considerable pressure on the ice from below, and when the ice cracks the water would surge up through the cracks at high speeds, and in the process it would go from 32°F to −270°F. When it hit the surface it would be "boiling hot" and would quickly vaporize. To us it would appear as a water geyser, shooting ice crystals and snow thousands of feet above the surface. This "snow" would extend out for 30 miles or more around the cracks.

As we look at the images from *Voyager* and *Galileo* we see that the surface is cracked, some of the cracks running for thousands of miles. Furthermore, they are surrounded on both sides by dark material that has little depth. It looks like contaminated snow. Perhaps the water beneath the surface is dark—full of debris from the volcanoes, hydrocarbons, and perhaps life forms. Interestingly, many of the dark lines have a bright stripe down the middle. What causes it? We cannot be certain, but it may be due to later surges of cleaner water.

In addition to the long cracks there are ice flows and huge icebergs up to eight miles across on the surface. In places the ice appears to have

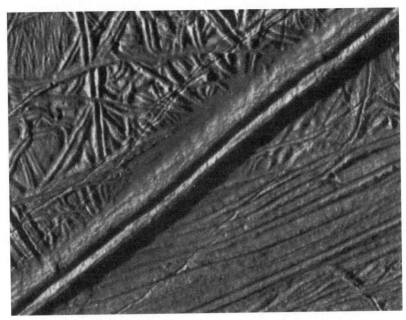

Close-up of large crack in Europa's surface. (Courtesy NASA)

broken into pieces that have shifted away from one another. In some cases the pieces are rotated and twisted at strange angles. This could be the result of a form of tectonics on the surface, similar to that on Earth where large plates shift around causing earthquakes. In this case, however, it would be due to warm currents from subsurface volcanoes or other eruptions on the ocean floor hitting the underside of the ice.

Another interesting feature of Europa's surface is that it is relatively free of craters, indicating that it is a young surface. One of the few visible craters is called Pwyll. The crater itself is 18 miles in diameter; across it and extending out to 27 miles is a red–brown region. It appears to be the remnants of liquid from the interior of the moon that splashed out when the meteorite struck. If so, the water may contain considerable organic matter.

The scenario I have presented is fascinating, but much of it is still speculative. We would like to be certain; we would like to see if there is water beneath the ice. And, indeed, there is a proposed mission to Europa: *Europa Ice Clipper* will be launched in December 2001. It will reach Europa in June 2009 and is scheduled to return a sample to Earth in 2014. It is not a manned mission so no one will be there to drill through to the water, but plans are on the drawing board for an ice-penetrating robot. The four-and-a-half-foot robot will carry a tethered vessel that will come to the surface after it strikes water. From there it will radio back information to Earth about the chemical content of the water. Scientists are looking forward eagerly to the mission.

Titan

Our other prospect for life in the solar system is Titan, the largest moon of Saturn. The life in this case wouldn't be beneath the surface, for Titan doesn't have an ocean of water either on its surface or beneath it (as far as we know). But we know it has an extensive atmosphere, and one of the most important discoveries of *Voyager* was that this atmosphere is strangely Earthlike. In other words, like the Earth's atmosphere it has considerable nitrogen.

Titan is larger than both Mercury and Pluto, and its atmosphere is more dense than that of either Earth or Mars. Like Europa it may have gone through a slightly balmy period early on, but Titan is over twice as far from Saturn as Europa is from Jupiter (and, of course, Jupiter is more massive) so this brief period likely had little effect on it.

Titan was, at first, a disappointment to the *Voyager* scientists. As the only moon in the solar system with an extensive atmosphere, it generated considerable excitement as *Voyager* approached it. What would the surface be like? Methane had been identified in the spectra many years earlier and most astronomers thought that there would be a large ocean of methane, making it the only object besides Earth that had liquid on its surface. But as *Voyager*'s cameras focused on the moon the atmosphere was found to be opaque; it was dense and orange, and nothing could be seen through it. Fortunately, there were several other instruments aboard that gave us considerable information. The atmosphere contained nitrogen in all three forms: molecular nitrogen, ionized nitrogen (with lost electrons), and non-ionized nitrogen. The pressure of the atmosphere was high—half again as great as Earth's atmosphere—and the surface temperature was close to what was expected, $-290°F$.

There had been controversy about Titan's size. Was it the largest moon in the solar system? Its radius turned out to be 2575 kilometers (1600

Titan. (Courtesy NASA)

miles), making it slightly smaller than Ganymede, the largest moon of Jupiter.

What was the surface of Titan like? Methane can exist in all three states near the surface, namely, as a gas, liquid, or solid. At the surface it would be liquid but just above, it would be a gas. It seems likely that there are methane clouds in the lower atmosphere, and methane rainfall onto the surface. But does it fall into a methane ocean? That now seems less likely than it did a few years ago. In 1992 a group led by Dewey Muhleman of JPL bounced a radar beam off Titan. They found that the surface does not reflect radar as an ocean would, and concluded that if there is liquid on the surface it likely has a patchy distribution—small seas and lakes. Furthermore, it now appears more likely that the liquid is ethane. If there are seas and lakes of ethane, however, there will be considerable methane and other impurities dissolved in them.

Although the most common gas in the atmosphere of Titan is nitrogen, many other gases are present. Argon, for example, makes up about 12 percent, and methane, hydrogen, ethane, propane, and other gases are also present. Most important, though, free oxygen is not present, and Titan's atmosphere is therefore reducing. We know that life on Earth arose in a reducing atmosphere. Titan's atmosphere is therefore much like the early Earth's, and we may be able to learn a lot about the early Earth by studying it.

The close-ups of Titan by *Voyager* showed us nothing of the surface, but Peter Smith and his team at the University of Arizona have been able to penetrate its atmosphere. They used the Hubble wide-field planetary camera to look at Titan in the infrared. Titan's atmosphere is partially transparent at these wavelengths. They snapped 50 images of the moon, then processed them to remove the varying effects of the atmosphere. Two main features were visible in what remained: a bright region about the size of Australia and a mottled dark region. Smith is not sure what either region is, but he presumes the bright spot may be a large mountain or high continent, lighter than the surrounding regions because it has been washed free of the dark hydrocarbon goo that is believed to cover much of the moon.

This and other mysteries about the surface may soon be cleared up. A spacecraft called *Cassini* was launched in October 1997 from Cape Canaveral. After several orbits around the inner solar system it will fly outward to Jupiter and Saturn, reaching Saturn in July 2004, and then moving on to Titan. Aboard *Cassini* is a small probe called *Huygens* that will be dropped through Titan's atmosphere and land on the surface, where it will take several measurements.

What is it likely to see? Will the surface be shrouded in an orange fog? Experts on Titan tell us that this is unlikely. The horizontal visibility should be relatively good. The light will be dim since Titan is ten times further from the sun than Earth, but the cameras of *Huygens* should be able to penetrate the light. It will be like a bright, full-moon night here on Earth.

Other Moons

Of the other moons of Saturn, only one called Iapetus is of much interest in relation to life. Although it is unlikely that it contains life, it does

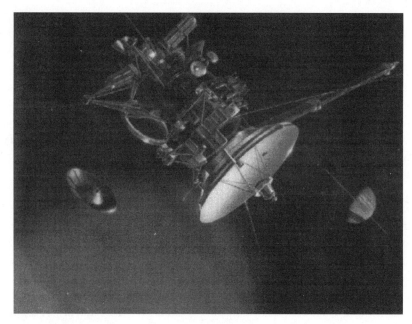

Artist's conception of Cassini. (Courtesy NASA)

have a puzzling feature: as it spins on its axis its brightness changes by a factor of seven. In other words, one side of it is bright, the other side dark. We have no idea why this is the case. It could be that the brightness arises from frost. *Cassini* will likely get a good view of Iapetus as it passes on its way to Titan.

Jupiter's two largest moons, Ganymede and Callisto, are also of some interest. Both are believed to have liquid water beneath the surface, but the prospect for life on either of them is much less than for Europa.

Triton, the largest moon of Neptune, is also of some interest. *Voyager II* discovered a tenuous, nitrogen-dominated atmosphere that has about ¹⁄₁₆ millionth the pressure of our atmosphere. Frozen nitrogen was also seen on the surface, and there is evidence of carbon dioxide and carbon monoxide. Of particular interest are the active geysers or jets that shoot material above the surface, causing plumes that extend out hundreds of miles from their source. But the surface, at −236°C (−392°F) is extremely inhospitable to life. It is the coldest sunlit surface in the solar system—colder even than Pluto's surface (which also eliminates Pluto as a serious candidate for life).

Venus

Turning now to the planets, let's begin with Venus. The surface of Venus is so hot it's unlikely that any form of life could survive there. But as you move above the surface, temperatures decrease, and there is a region about 40 miles up where temperatures are moderate—about room temperature—and the pressure is about the same as a high mountain site here on Earth.

This might seem like a good place for life. But when you look closer you see serious problems. Any life in this region would be floating in sulfuric acid, and it is unlikely that any form of life could survive for long in it. Furthermore, there is little water vapor in the atmosphere, and life needs water. Also, Venus has no ozone layer, so any life in the region would need protection from ultraviolet radiation. Still, it might have a protective shell, and other features we are not familiar with, so—despite the odds—it is possible that some form of life is present.

What is much more likely, however, is that Venus had life in the past. With the high temperature of the surface, however, it would be difficult to find any fossil remains. Venus is closer to the sun than Earth and even without a strong greenhouse effect it would be hotter. But early on the sun was less luminous than it is today—about 25 percent less—and Venus would have been cooler. Life may have formed during this period, but it wouldn't have lasted long.

Jupiter

Of the four gas giants, the only one worth considering in relation to life is Jupiter. Jupiter doesn't have a solid surface as the terrestrial planets do; the atmosphere just thickens and becomes more dense until we are finally in a region of liquid hydrogen and helium. But like Venus it has a region in the atmosphere where moderate temperatures prevail, and in this case there is no sulfuric acid. Life there would, of course, have to be airborne life—spores or perhaps tiny gas-bag creatures.

It is even possible that large gas-bag creatures exist that feed on the smaller ones and on the spores that float in the atmosphere. These creatures would likely use hydrogen to maintain their position, warming and cooling it to change to a different level. Jupiter is the most likely of all the gas giants to have such creatures, as it is the most chemically active.

Of particular importance, Jupiter has all the gases of an ideal primitive atmosphere: methane, ammonia, hydrogen, and water vapor. In addition, it has a variety of other gases that are important in relation to life, gases such as hydrogen cyanide and carbon monoxide. Furthermore, lightning is common in the outer clouds and would serve as an energy source. On the basis of this, it is quite possible that life may have formed, but it would no doubt be difficult to find evidence of it because of the turbulence that exists in Jupiter's atmosphere.

Asteroids, Meteorites, and Comets

In chapter 6, we discussed the meteorite that fell in the Antarctic that we now believe is from Mars. Meteorite falls are common on Earth and many have been shown to contain amino acids and other complex biological molecules; even the components of DNA have been found.

Meteorites are generally classified as stones, irons, or stony-irons, according to their makeup. A class that has been of particular interest is called carbonaceous chondrites, which contain a large amount of carbon and carbon compounds, along with nitrogen and water.

One of the major difficulties in analyzing meteorites is contamination. The longer they lie on the Earth after they strike it, the greater the chances of contamination. It is essential, therefore, that we locate the meteorite as soon after it falls as possible. A carbonaceous chondrite was located within a day of its fall near Murcheson, Australia, in 1972. It is now referred to as the Murcheson meteorite, and analysis of it has shown that it contains many of the basic molecules of life. Seventy-four different amino acids were identified in it, eight of which occur in proteins on Earth, and eleven that play a part in terrestrial biology. How can we be certain that they aren't a result of contamination? Most of the amino acids that occur in it do not occur on Earth; furthermore, all of the Earth's amino acids have a left-hand helical structure (like a left-hand corkscrew), and many of those found in the Murcheson meteorite are right-handed. Some of the building blocks of DNA and RNA, namely, guanine, adenine, uracil, cytosine, and thymine were also found in it.

The Murcheson meteorite is about 4.5 billion years old and spent almost its entire life in space. This means that the basic molecules of life can be produced even in the cold depths of space. The warm, early conditions of Earth may not be needed.

We have not been able to analyze comets in the lab as we have meteorites, but we can look at their spectra, and it appears that many of the basic molecules of life are contained in them. In fact, as we suggested earlier, it may have been comets colliding with Earth that gave us some of the first life molecules.

The discovery that there may be life on Mars and perhaps on Europa or Titan is encouraging. To many, however, it may be a disappointment that at most it will only be a primitive form of life. For more advanced forms we have to look to the stars.

chapter 9

Life on Planets beyond the Solar System

Having examined the solar system, we know that, aside from Earth, none of the objects in it are likely to harbor a higher form of life. As we move out beyond the solar system, out into the stars, however, we find the prospects for more complex forms of life—even a civilization such as ours—are much greater. Our galaxy contains an uncountable number of stars, somewhere in the neighborhood of 100 billion. That's roughly the number of grains of sand on all the beaches in the world. Most of these stars are not going to be good candidates as life-supporting systems, but if only a tiny fraction, say 1 percent, are, we would have a lot of life.

Even though the chances are obviously high that there is alien life out there somewhere, the difficulties of detecting it are overwhelming. We therefore have to ask ourselves what we would look for. First, we would need an ideal star, a star similar to our sun since we know there is life around it. But what about other stars, other types of stars? If we lined up all the stars starting with the least massive, we would find that they range from tiny red dwarfs, about 80 times as massive as Jupiter, up to huge blue giants, over 80 times as massive as our sun. We would find a few red giants even larger than the most massive stars, but they are bloated stars—huge because their surfaces are extended far beyond their usual limit.

In the early part of this century two astronomers, Ejnar Hertzsprung of Denmark and Henry Norris Russell of Princeton University, independently plotted a graph of absolute brightness (magnitude) versus surface temperature for a large number of nearby stars. Most of the stars fell on a diagonal across the graph, but a few fell in the upper-right-hand region of the graph and a few in the lower-left-hand region. Not surprisingly, this is what you would get if you plotted height versus weight for a large group of people. Most people are of average weight for their height and would lie along a diagonal across the graph, depending on their height. A few, however, would lie off the diagonal because they are too heavy or too light for their height. Stars are like people in this respect.

The diagram drawn by Hertzsprung and Russell has been named in their honor; it is now called the HR diagram. The diagonal across the center is called the main sequence. Our sun, being an average-sized star, lies roughly in the middle of this diagonal. Near the top of it are huge blue giants, and near the bottom, red dwarfs. It will be useful to refer to this diagram in our discussion.

One last point before we consider which of the stars in this diagram are best suited for life. Stars of various temperatures are also grouped into what are called spectral types. These types are referred to by the letters O, B, A, F, G, K, M, and N, with the hottest stars being the O types (blue giants) and the coolest, the N types (red dwarfs). Surface temperature decreases uniformly between these two types. This means that instead of

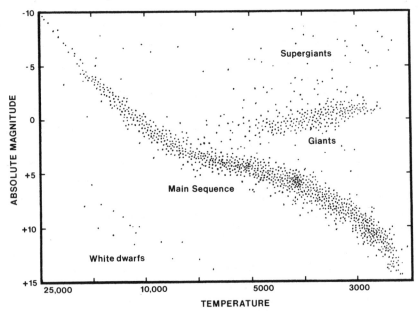

The H. R. Diagram. O, B, A, F, G stars are to the left in the diagram; K, M, N are to the right.

temperature on the horizontal axis of the HR diagram, we can plot spectral type. Our sun, incidentally, is a G-type star with a surface temperature of approximately 10,000°F.

Our sun is obviously an excellent candidate for supporting life, since we know it supports Earth. But what about other spectral types? As we look down the main sequence at the stars we see there are problems with many of them. The stars at the top of the main sequence, namely, the large blue giants, are the shortest-lived stars of the universe. They produce heavy elements and explode, distributing them to space where new sys-tems are formed from them, but they do this in a relatively short time (astronomically speaking). They live only about 50 million to 100 million years, which is much too short for life to form. As we saw in chapter 3 this takes billions of years. So, although they play a critical role in the formation of life by creating heavy elements, they are not good candidates for sup-porting life. On the basis of this we have to eliminate all O, B, and A stars and perhaps some of the F types (each of the spectral classes is subdivided from 0 to 9).

Turning now to the bottom of the main sequence we find the K, M, and N types, in other words, the dim stars and red dwarfs. The problem with these stars centers around their life zone. The life zone, or ecosphere, of a star is the region where water is in a liquid form most of the time. Since life needs liquid water, it is the region where conditions are suitable for life. The ecosphere of the sun extends roughly from the orbit of Venus to that of Mars. Earth is in the center with Venus and Mars on the outer edges.

Large stars have large ecospheres: small stars have small ones. This means that red dwarfs will have relatively narrow ecospheres that are quite close to them. Two problems can arise because of this. We know that most planets trace out elliptical orbits, and it is difficult for an elliptical orbit (because of its elongation) to remain inside a narrow ecosphere. More serious than this, however, is that even if it did, the planet would eventu-ally become "tide-locked" to its star, in the way our moon is tide-locked to us. In other words, the same side would always point toward the star, and this would have disastrous consequences: temperatures would likely be extreme—too hot or too cold. Because of this we have to eliminate all M and N stars and most Ks. This leaves us with only a small range around stars similar to the sun that are ideal candidates.

Even if we have an ideal star with a planet within its ecosphere, however, other things have to be considered. Water, for example, has to be available. As we saw earlier, it is by far the best solvent for life, and it is

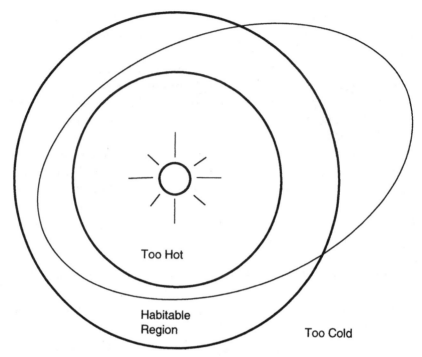

The habitable or life zone around a star. An elliptical orbit is shown passing in and out of zone.

difficult to imagine life forms that use other liquids. The atmosphere of the planet would have to be satisfactory; it need not be the same as ours, but oxygen would likely be necessary. And finally there would have to be protection from radiation. We are protected from ultraviolet radiation from the sun, for example, by the ozone in our atmosphere.

Several other factors are also important. If the planet is too massive, it's unlikely it would lose the huge hydrogen–helium atmosphere that formed around it early on. It would be like Jupiter or Saturn, which we know are unsatisfactory for most forms of life. Furthermore, if its mass was too small it would have insufficient gravity to retain an atmosphere, as in the case of Mercury. Surprisingly, the presence of a moon is also important. Our moon helps stabilize the orientation of the Earth's spin axis. The orientation of Mars's spin axis, for example, has changed considerably over the years because it doesn't have a stabilizing moon. Many scientists believe that changes such as this are responsible for ice ages. Our moon was also likely important in relation to the evolution of life in that it produced tides that allowed life to migrate from the oceans to land.

The Nearby Candidates

Now that we have some idea of the type of stars we are looking for, let's consider some of the nearby stars. Nearby stars are of most interest because they are the only ones we would have any chance of communicating with if they did harbor an advanced form of life. Radio signals travel at the speed of light, so it would take 4.3 years for a signal to reach our nearest neighbor, Alpha Centauri, and another 4.3 years to receive a reply if there were a civilization there that understood it. Most good candidates, on the other hand, are likely to be much farther away than this.

Let's begin by considering the nearest two dozen stars. Most of these stars are smaller and less luminous than our sun; only three are more luminous: Sirius, Procyon, and Alpha Centauri A. And the first two of these are not good candidates; both are too young. Life forms, except for the most elementary, took approximately 3 billion years to develop here on Earth, and both Sirius and Procyon are younger than this.

Of the stars on our list, in fact, only four are reasonably good candidates: Alpha Centauri A and B, Tau Ceti, and Epsilon Eridani. We usually think of multiple star systems as poor candidates because the ecospheres around them would be continuously changing as the two stars revolved around one another. Alpha Centauri A and B, however, are separated by a relatively large distance and planets could support life if they were sufficiently close to one of the stars. Alpha Centauri A has a mass about 10 percent greater than our sun and a luminosity about 50 percent greater, so if there were a planet within its ecosphere there is a chance it could harbor life. Alpha Centauri B, on the other hand, is only about half as luminous as our sun and much less likely to harbor life.

Tau Ceti is, perhaps, the best nearby candidate. Like the sun, it is a single G-type star, but it only has about half the sun's luminosity. The next best candidate is Epsilon Eridani. It is a K-type star and only about 30 percent as luminous as the sun, so its ecosphere would be relatively narrow. A few others within this group are possibilities, but the probability that they have planets that harbor life is small.

Early Searches for Extrasolar Planets

If we are to find life we must first find a planet, so let's direct our attention to the search for planets around some of the nearby stars. Much of the early interest in extrasolar planets centered on the dim star known as Barnard's star. It's only six light-years away, but it is so dim we cannot see it with the naked eye. Yet, strangely, it is our second-nearest neighbor; only the triple system Alpha Centauri is closer. It's a red dwarf, five times less massive than our sun and about half as hot.

Barnard's star was discovered by Edward Barnard of Lick Observatory in 1916. Comparing a photograph of a section of the sky in the constellation Ophiuchus to a photograph that had been taken 22 years earlier he was surprised to find that one of the stars had moved considerably. He measured its proper motion (angular velocity across the sky) and was amazed to find it was the largest ever recorded. The star obviously had to be close.

Peter van de Kamp first took notice of the star in 1938, shortly after he became director of the Sproul Observatory at Swarthmore College near Philadelphia. One of his first programs after assuming the directorship was a search of nearby stars for planetary companions. He knew there was little chance of detecting planets directly, but a method for detecting them indirectly had been used by Friedrich Bessel of the Konigsberg Observatory in Prussia in the mid-1800s to show that Sirius had a tiny companion.

Van de Kamp knew that if a star had a planet, the star would "wobble" as it moved across the sky. To see why, consider Jupiter and our sun. It might seem that Jupiter revolves around the sun, but this isn't so; the two objects actually revolve around their center of mass. This is the point where they would balance if Jupiter was on one end of a teeter-totter and the sun was on the other end. The sun is thousands of times heavier than Jupiter, so the balance point is obviously going to be close to the sun; in fact, it's just outside its surface.

This means that when a star and its planet move through space they revolve around their center of mass. At the same time, however, they trace out a much larger orbit around our galaxy. Looking at the star through a telescope, we wouldn't be able to see the planet, but we would see the star "wobble" as it moved through space. This wobble is a clear signature of the planet's presence. The effect is extremely small, and to see it you have to observe the path of the star over a time of several orbital periods, which is usually many years.

Van de Kamp began photographing Barnard's star in 1938. Two or three times a month he and several colleagues would expose a dozen or more plates, then plot the position of the star carefully on a graph. Surprisingly, within a few years there was a small deviation from a smooth curve, but not enough for van de Kamp to be sure. The period appeared to be 24 years, so he knew he would have to collect data at least this long to be sure. Finally, in 1963 he was convinced there was enough evidence to announce his findings. According to his measurements there was a planet about 1.6 times as massive as Jupiter orbiting Barnard's star in a highly elliptical orbit.

The announcement caused a flurry of excitement. No one had ever detected a planet beyond the solar system before. But van de Kamp wasn't finished; he continued observing the star and in 1968 he announced that his data indicated an object 1.7 times as massive as Jupiter was orbiting with a period of 25 years. The orbit was even more eccentric than he had previously believed. A year later he announced that an even better interpretation of his data was two planets in orbit around the star. One had a mass approximately 80 percent that of Jupiter and an orbital period of 12 years; the other was 10 percent more massive and had a period of 26 years.

If van de Kamp's discovery was verified by other astronomers, it would be a momentous discovery. The implications were enormous. Barnard's star was one of the nearest stars in the sky and if it had two planets, most other stars likely had planets.

What were the planets like? Did they contain life? Astronomers could hardly contain their excitement. It was soon obvious, however, that these planets, assuming they existed, were poor candidates for life. Barnard's star was a red dwarf and as such was composed mostly of hydrogen and helium. Such a small star would not likely have rocky planets orbiting it. Its planets, if indeed they were planets, would likely be gas giants like Jupiter.

Several astronomers began checking to see if they could verify van de Kamp's results. Robert Harrington, a former student of his who was now at the United States Naval Observatory, looked carefully at the data. He compared the curve that van de Kamp had obtained to other data that had been taken at Sproul Observatory and found there was a strange similarity. John Hershey of Sproul Observatory also looked at several stars measured by van de Kamp and his colleagues. They all had a significant deviation in 1949 and again in 1957. Checking the records he found that the telescope had been disassembled and cleaned on these dates.

George Gatewood of Allegheny Observatory also gathered photos from several observatories that had been taken over the years. Using them he plotted the position of Barnard's star over several years, and although he found a small deviation from a smooth curve, he found nothing to indicate there were two Jupiter-sized planets orbiting it. Gatewood then started his own observing program, and after a few years found that his data contradicted that of van de Kamp.

Despite the evidence against his interpretation, van de Kamp did not give up. He continued taking data and in 1978 announced new findings. Again he found two bodies in orbit around Barnard's star, but now they had 0.8 and 0.4 Jupiter masses, with respective orbital periods of 11.7 years and 20 years. Three years later he revised them again slightly.

As time went on van de Kamp's data came under greater and greater scrutiny and eventually there appeared to be too many problems for anyone to take the data seriously. To his death, however, van de Kamp was certain his interpretation was correct.

A New Technique

One of the major difficulties in measuring the exact position of a star is scintillation of its light. The path that the light takes as it passes through our atmosphere changes slightly due to small changes in the atmosphere (e.g., hot air rises). We see these changes as the twinkling, or scintillation, of the star, and it causes a blurring of the image on a photographic plate. The beam striking the plate "wobbles" slightly as it "burns" its image into the plate. One way of getting around this is what is called the "speckle technique." A series of short exposures is taken—each only 10 to 20 milliseconds in length. During such a short period the beam has little time to wander. A large number of short exposures such as this are taken, then they are superimposed. The resulting image is much sharper than it would have been if a single long exposure had been taken.

If, in addition to this, the infrared is used, the technique is of even greater value. Infrared is heat energy and the most likely type to be emitted by a planet. The radiation we detect from a planet is reflected from its surface, and most of it is infrared. The problem in trying to detect this radiation is that it is overwhelmed by the radiation from its parent star. The visible light from a star like our sun is a billion times greater than that from a planet like Jupiter, and the chances of detecting a Jupiter-sized planet in the visible are therefore extremely small. Planets, on the other hand, emit considerable more radiation in the infrared than they do in the visible, while most stars radiate less. The ratio of brightness between the star and the planet therefore can be reduced as much as a thousand in the infrared. While the planet is still overwhelmed by the star's infrared, there is a much better chance of detecting it.

Donald McCarthy, Jr. and Frank Low of the University of Arizona and Ronald Probst of the National Optical Astronomy Observatories (NOAO) developed a technique of this type in 1984. Within a short time McCarthy and his team's attention had been caught by a star called Van Biesbroeck 8 (VB8) named for the Belgian-American astronomer George Van Biesbroeck, who had studied it earlier. Using the four-meter telescope at Kitt Peak they obtained data that indicated that this star had a faint companion. They repeated their observations over several nights with two different telescopes and were finally convinced that a tiny object was orbiting VB8, its light 40 times fainter than the star.

The announcement of the result caused considerable excitement. It was sensationalized in the press as the "first planet discovered outside of the solar system" and there was a lot of speculation about it. But when other astronomers started looking for it to verify the discovery they couldn't find it. And surprisingly when McCarthy and his team looked for it again to prove that it indeed existed, they couldn't find it either. Where had it disappeared to? McCarthy had no answer. Most astronomers eventually became convinced that it was a case of mistaken identity: It never existed in the first place.

IRAS and Beta Pictoris

It wasn't long, however, before there was further evidence for extrasolar systems. This time it came not in the form of a planet, but as a disk of debris. The first inkling that there might be disks around nearby stars came with the launch of the IRAS satellite in 1983. In its relatively short lifetime (less than a year), IRAS mapped the entire sky in the infrared. Thousands of infrared sources were discovered. Many, of course, were dim red stars, but some were different and at first, controversial. The first of these new types of sources was, strangely enough, the bright star Vega, one of the brightest stars in the sky. The data indicated that Vega had an excess of infrared—far more than a star of this type, age, and size should have. What was this "infrared excess?" H. Aumann of JPL and Fred Gillett of the Kitt Peak Observatory carefully measured the emitting region. It was huge, like a disk, and it had a temperature of −185°C. Vega appeared to be surrounded by a gigantic flattened disk—an accretion disk. Most of the particles, according to their estimates, were the size of grains. Because Vega was a relatively young star—only about 300 million years old—compared to the sun's 5 billion, many astronomers became convinced that we were seeing a solar system in formation. According to theory, our system formed from a similar accretion disk around the sun.

The announcement caused considerable excitement and astronomers soon began looking carefully through the data gathered on other systems. Before long several other similar sources were found.

Two astronomers, Brad Smith of the University of Arizona and Rich Terrile of JPL, saw the new discovery as an opportunity. They were headed for the Las Campanas Observatory in Chile on a different project, but they decided to take a look at some of the excess infrared sources. Using a 100-inch telescope, they equipped it with a coronograph—a tiny mask that cuts off the light of the star—and looked at several of the candidates. One of them was Beta Pictoris, a star about 53 light-years away.

After four nights of observations they took their data back to their respective institutions for processing. To their surprise and delight they found streaks of light extending out from Beta Pictoris in two directions. A disk 30 times larger than our solar system, and considerably larger than the disk of Vega, was clearly visible. It was the first time a disk had actually been observed. Astronomers are now convinced it consists of small ice crystals and grains of silicate and perhaps carbon compounds.

Another surprise came when they examined the disk in detail. It did not extend all the way up to the star. The 1.5 billion miles closest to the star were clear of debris (this corresponds to the distance in our system from the sun out to Uranus). Astronomers have speculated that this region may be clear because the debris that was once here has accreted into planets.

Dana Backman of NOAO continued studying the data from IRAS. Within a few years he had identified 25 stars with excess infrared, two of which had a gap between the inner edge of the disk and the star.

Large Planet or Brown Dwarf?

Backman continued his quest for excess infrared sources by initiating his own search in 1987. He wanted to find out how common the phenomenon was. Within a short time he found that 10 percent to 20 percent of all young stars appeared to be circled by disks of matter. Furthermore, during his research he discovered tiny points of light close to the stars Gliese 803 and 879. Backman was not sure what the objects were; they were dim, but they could be tiny stars.

Strangely, a similar type of object had been found about two years earlier by William Forrest of the University of Rochester. In July 1985, Forrest was using the infrared telescope (IRTF) on Mauna Kea to look at the faint red star Gliese 569 when he came upon a tiny point of light beside it. At first he thought it was a tiny dim star. Then Ben Zuckerman of UCLA and Eric Becklin of the University of Hawaii came upon a tiny infrared excess close to the star Gielas 29–38. And a little later Zuckerman came upon a similar object near the star GD165.

What were these objects? They weren't small enough to be planets, yet they seemed to be too small to be stars. For years astronomers had been interested in this region—the region between planets and stars. We know that the smallest star has a mass about 80 times that of Jupiter. If a mass of gas less than this condenses, the core doesn't reach a high enough temperature to trigger nuclear reactions and it doesn't become a star. And yet internal compression heats the gas and the object shines dimly as a result of this heat. Astronomers have named such objects brown dwarfs. According to estimates they should have a mass between about 20 Jupiter masses and 80 Jupiter masses.

How, then, do we distinguish between a small brown dwarf and a large planet such as Jupiter? As it turns out it is difficult, but there is a significant difference between the two objects. Brown dwarfs condense out of a cloud of gas, and are, with the exception of the nuclear core, similar to stars. In other words, they are made up mostly of hydrogen and helium, and have a gaseous core. Planets, on the other hand, form in an accretion disk around a star. Some of these planets can be large—perhaps more massive than the smallest brown dwarfs—but they are formed by accretion and have a rocky core as Jupiter and Saturn do.

Could we detect the difference between these two objects if we could get their spectra? As it turns out, it would be difficult. Both a large Jupiter-like object and a brown dwarf are composed of hydrogen and helium near the surface, and their spectra would be similar.

Most of the objects discussed in this section were assumed to be brown dwarfs, but there was considerable uncertainty. And as we will see this uncertainty still plagues astronomers today.

Doppler Shifts

We have seen that there are two methods of detecting planets in orbit around nearby stars: direct and indirect. The method used by van de Kamp is indirect; we do not see the planet directly but are able to predict its presence by its effect on its parent star. In the direct method we would detect light from the planet. The discoveries mentioned earlier—tiny visible companions—were made as a result of direct measurements, in other words, measurements of light from the object itself.

There is another indirect method that is based on spectra. If a planet

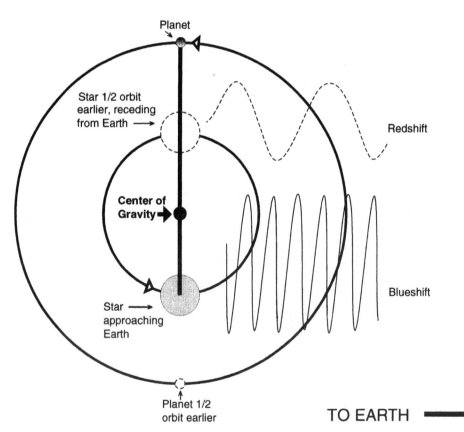

The Doppler effect. Star receding from Earth is redshifted. Star approaching Earth is blueshifted.

orbits a star, the star is going to "wobble," as we have seen. This means at some point the star will be moving toward us and at other points it will be moving away. It will therefore exhibit alternating blueshifts and redshifts in its spectrum. We can only get the radial component of this motion and therefore cannot be sure of the direction of the plane of the orbit. Because of this it is difficult to estimate the mass of the companion; we can only get the minimum mass and estimate from it what the true mass is. Nevertheless, it's still an excellent technique for determining the presence of planetary objects, and most of the discoveries in recent years have been made using this method.

One of the first to use the method successfully was a Canadian team consisting of Bruce Campbell, Gordon Walker, and Stephenson Yang. The major difficulty prior to their work was obtaining an accurate estimate of the star's velocity. Standard Doppler measurements only gave accuracies of about 2000 miles/hr. An accuracy of about 60 miles/hr was required to detect a Jupiter-sized object.

A set of reference lines was needed to measure the star's spectra accurately. These reference lines were usually supplied by a separate source—an iron arc lamp, for example. Campbell and his colleagues decided to superimpose the spectrum of the reference source directly on the star's spectrum. If the light from the star passed through a filter that had its own characteristic spectral lines, these lines would be imprinted on the star's spectrum, and accurate measurements could be made.

Campbell selected hydrogen fluoride as the filtering gas. It is a noxious and highly corrosive gas so it had to be handled carefully. It was put in a specifically designed plexiglass container with sapphire windows. With this device they were able to get their accuracy down to 30 miles/hr—easily enough to detect a Jupiter-sized planet.

Using the 144-inch telescope of the Canada–France–Hawaii Observatory on Mauna Kea they began a search in 1980 of approximately two dozen nearby sunlike stars. For eight years Campbell and his group measured the spectra of these stars, then fed the data into a computer program to determine the movement of each star. In 1988 they announced their findings: nearly half the stars they were studying had small variations. Of these, one called Gamma Cephei was of particular interest. According to their results a planet about 1.7 times as massive as Jupiter, with a period of 2.7 years, was orbiting the star.

By 1992, however, they were having doubts. They could measure velocities of about 30 miles/hr, but material on the surface of the star can move upward, or downward, at speeds much in excess of this, and it is difficult to distinguish it from the overall velocity of the star.

A similar program was started by William Cochran of the University of Texas. Instead of hydrogen fluoride, however, he used iodine as his reference gas. Cochran found some interesting results for the star HD 114762; it appeared to have a dark companion but he couldn't determine its mass.

Planets at Last!

In October 1995, Swiss astronomers Michael Major and Didier Queloz of the Geneva Observatory announced the discovery of an object orbiting 51 Pegasi in the constellation Pegasis. It was Jupiter-sized but only ⅟₁₀₀ the Jupiter–sun distance from its star. The scientific world was shocked. Also shocked were two California astronomers, Geoffrey Marcy and Paul Butler of San Francisco State University. They had been working on a program to detect planets around nearby stars since 1987 and were sure they had such

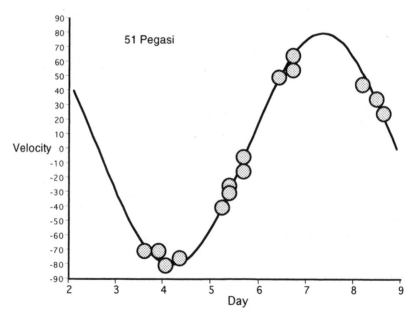

The light curve of 51 Pegasi.

Geoffrey Marcy and Paul Butler. (Butler is to the left.) (Courtesy San Francisco State University)

a head start that no one could scoop them. They were examining 120 nearby stars and were about halfway through their program when they heard the announcement.

Butler and Marcy rushed to their telescope at Lick Observatory to see if they could verify, or refute, the discovery. And lo and behold, after four days and nights of intense work they confirmed that 51 Pegasi *did* have a planet. It was a giant planet with half the mass of Jupiter and an orbital period of 4.2 days. As for 51 Pegasi itself, it was a star about 45 light-years away that was almost identical to our sun, so it was obviously a good candidate as a life-supporting system. But 51 Pegasi B, as the planet is called, is not a likely candidate for life. It is so close to its star that its surface temperature would be almost 2000°F—far too hot for any form of life.

Marcy and Butler knew that they had to process their data from the 60 stars they had measured. And they had to do it quickly. Others were close on their tails, and they could be scooped again. They begged and borrowed computer time, keeping six computers going day and night to analyze their data. One or the other of the two checked the computer's progress each morning. On the morning of December 30 it was Butler's turn. Looking at the numbers that had been crunched during the night he was stunned. "I was blown away," he said. One of the stars, 70 Virginis in the constellation Virgo, had the characteristic wobble of a star with an orbiting planet. Within days the two were confident they had a planet. The

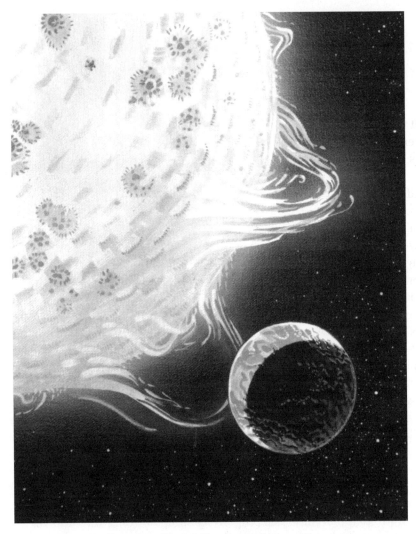

Artist's conception of Jupiterlike planet orbiting a distant star.

star 70 Virginis was also sunlike, at a distance of approximately 60 light-years. At 6.5 Jupiter masses, its planet was considerably larger than that of 51 Pegasi. It was 0.43 astronomical units [an astronomical unit (AU) is the distance from the earth to the sun] from its star and it had an orbital period of 117 days.

Particularly surprising was that it had an extremely eccentric orbit; in other words, its orbit was very egg-shaped. The temperature of this planet was far more interesting than that of 51 Pegasi. The top of the clouds were calculated to have a temperature of about 185°F, approximately the temperature of bathwater, and organic molecules could easily form and survive here.

Within a short time Marcy and Butler discovered a second planet, this one orbiting the star 47 Ursae Majoris near the Big Dipper. The planet 47 Ursae Majoris B has a mass of 3.5 Jupiters and a period of three years, and it orbits at a distance of 2.1 AU from its star. Its orbit is circular and the temperature at its surface is a relatively cool −80°C. Again the parent star is quite similar to our sun, with a mass roughly equal to it.

The California group now had two planets to announce to the scientific world and the logical place to do it was at the upcoming American Astronomical Society meeting in San Antonio; it was three weeks away. A silent tension filled the air as they waited to make the announcement, worried that the Swiss team might beat them to it. But they didn't, and again the news amazed the scientific world. Many felt that the first discovery was just a fluke, but now there were three—all of them large, Jupiter-sized planets. With the technology available only Jupiter-sized planets can be detected. None of them were particularly good candidates for life. Nevertheless, the discovery was extremely important for another reason. If there were planets around some of the nearby stars it meant that most stars likely had planets around them. And even if they were Jupiter-sized and unlikely to harbor life, large planets are likely accompanied by small planets. Furthermore, these large planets could have moons. The gas giants of our system all have moons, and as we saw earlier some of them have conditions that are suitable for life.

Since their original discovery, Butler and Marcy have discovered four more planets. Again, all are Jupiter-sized. They include Rho Cancri B with a minimum mass of 0.84 Jupiters (84 percent that of Jupiter) and an orbital period of 14.7 days; Tau Bootis B with a minimum mass of 3.87 Jupiters and an orbital period of 3.3 days; Upsilon Andromedae B with a minimum mass of 0.68 Jupiters and an orbital period of 4.6 days; and finally, 16 Cygni B with a mass of 1.5 Jupiters and a period of 2.2 years. The last was discovered in conjunction with William Cochran and Artie Hatzes of the University of Texas.

The planets divide themselves into three main groups: hot Jupiters, cool Jupiters, and eccentric Jupiters. The hot Jupiters are those that orbit close to their parent star; they include 51 Pegasi B, Rho Cancri B, Tau Bootis B, and Upsilon Andromedae B. The cool Jupiters are the ones similar to our Jupiter—those that orbit about the same distance from their stars as Jupiter is from the sun. They include 47 Ursae Majoris B and Lalande 21185 (recently discovered by George Gatewood). Finally the eccentric Jupiters, those with highly eccentric orbits are 70 Virginis, 16 Cygnus B, and one called HD 114762, discovered by David Latham of Harvard–Smithsonian and several colleagues.

Needless to say, the discovery of these new planets has raised a lot of questions. Prior to the discoveries most theorists assumed that if planets were found they would be similar in size and distribution to the ones in our system. Large Jupiter-sized planets, according to accepted theories, cannot form close to their star for two reasons. Ice condensed far out in the solar system, roughly where Jupiter is, and as we have seen, it is the icy-core planets that grow large. Furthermore, when the central protostar becomes a star there is a solar gale that rushes out into the region around the star, clearing it of debris. A large gas giant could not survive this gale if it is close to the star; all its accumulated hydrogen and helium blow away.

This leaves us with a problem. How did these planets, which we assume are gas giants like Jupiter, form so close to their star? Or indeed did they form there? Most astronomers are now convinced that they couldn't have. But if they formed farther out, how did they end up there? Several theories have been put forward in recent months. Doug Lin and Peter Bodenheimer of the University of California at Santa Cruz and Derek Richardson of the Canadian Institute for Theoretical Physics showed that a planet can lose energy and begin spiraling toward its star. According to their calculations, it is saved from catastrophe when some of the star's angular momentum from its spin is transferred to the falling star.

Fred Rasio and Eric Ford of MIT showed that if two Jupiter-sized planets formed around a star, gravitational interactions between them could eject one from the system and push the other inward toward the parent star. It would settle in an orbit close to the star. A slight variation on this was put forward by Weidenschilling and Marzari in 1996. They showed that if three or more giant planets form around a star, their orbits

Some of the extrasolar planets that have been discovered so far compared with the inner solar system. MJ is the mass of Jupiter.

can become unstable as they gain mass via accretion from the circumstellar disk. This instability would likely lead to collisions that would eject some of the planets from the system and leave others closer to the star, frequently in highly eccentric orbits. Another possibility is that they form beyond five AU but slowly migrate inward as the disk dissipates. There are, indeed, several possibilities and they may force us to take a serious look at the accepted theory of the origin of our system.

Although most astronomers were excited about the discoveries, a few were skeptical. David Gray, an astronomer at the University of Western Ontario, published an article in *Nature* showing that the variations in the case of 51 Pegasi, Tau Bootis, and possibly others, were likely caused by rhythmic pulsations. Major and Queloz, the discoverers of 51 Pegasi, along with Marcy and Butler quickly countered Gray's arguments. They pointed out that if oscillations were occurring in Tau Bootis there would be corresponding brightness variations, and there weren't. They also questioned why stars that are all quite similar to our sun would "oscillate" at different frequencies, when our sun shows no oscillation whatsoever.

Recently another Jupiter-sized planet was discovered by Robert Noyes of the Harvard–Smithsonian Astrophysical Observatory and several colleagues. It is orbiting Rho Coronae Boreallis, a visible star that is about 50 light-years from Earth. The planet is about 23 million miles from the star, which is closer to it than Mercury is to the sun. Its surface temperature is therefore about 400° to 500°F, so there's not much chance there is life on it. Nevertheless, it may be accompanied by smaller planets further out. The main difference with this new planet is that its distance from its star is between that of previously discovered planets.

Other Programs

The huge Keck telescope on Mauna Kea is being used in a program call Toward Other Planetary Systems (TOPS). TOPS was formed in 1988 at a workshop on extrasolar planets sponsored by NASA. At the workshop a three-phase program was outlined and presented to NASA officials for consideration. In the first phase, called TOPS-0, a ground-based search would be made for candidates. This was to be followed by TOPS-1, a space-based look for candidates, and finally TOPS-2 was to be a more extensive look in the future using more elaborate spacecrafts and perhaps telescopes on the moon.

The first phase got a boost in 1991 when the Keck Foundation announced it would partially fund a second 10-meter telescope, identical to Keck I. The Keck Foundation would provide 80 percent of the funding. The Keck telescope seemed ideal for the first stage of the TOPS program and other solar system studies, so NASA picked up the tab for the remaining 20 percent in return for an equivalent amount of observing time.

Of particular importance in such a program is resolution. If a star is about 30 light-years away and has a planet about one AU from it, the resolution needed to distinguish the planet, assuming the glare from the star is eliminated, is roughly 0.1 arc seconds. This is presently beyond any of the telescopes on Mauna Kea. But the two Keck telescopes, which are 85 meters apart, will soon be hooked together to form an interferometer. They will then be like two sections of a giant mirror 85 meters in diameter. With the two telescopes we will get excellent resolution but only part of the image. For a more complete image other telescopes will be needed. Plans are under way for the construction of four smaller "outrigger" telescopes that can be moved. They will have diameters of 1.5 to 2 meters. Used in conjunction with the two Keck telescopes they will give much better coverage of the image. According to current plans they will be movable between 18 different positions. Signals from the larger telescopes along with the four outriggers will be routed to the basement of the Keck II telescope where they will be combined.

This system will be used with adaptive optics to give us an even more powerful tool in our search. Adaptive optics is a technique using lasers that compensates for instabilities in the atmosphere. The TOPS program is still under way in principle, but as we will see many of NASA's plans have recently been changed.

The Hubble telescope will also no doubt be used in the search for extrasolar planets. The defect in the telescope was corrected in December 1993. The first images from the repaired telescope were everything astronomers had hoped for. With these changes the Hubble telescope has approximately ten times the resolution of one of the Keck telescopes (without interferometry or adaptive optics). In 1997 Hubble was equipped with a new infrared camera that will no doubt be particularly helpful in the search.

Beyond Hubble is a telescope that will be launched early in the next decade. Known as SIRTF (Space Infrared Telescope Facility), it will have a mirror 0.85 meters in diameter and will be able to detect infrared sources thousands of times fainter than those detected by IRAS. It is expected to remain in orbit about four years.

Direct Detection of Planets

Indirect detection of planets has been of tremendous value to us and will no doubt continue to be the major method used in our search over the next few years. But eventually we would like to detect planets directly; in other words, detect the radiation from the planets themselves. Indeed, this is the only way we'll be able to study them in detail. With indirect methods we are measuring properties of the parent star, not the planet, and the only thing that we can determine is the planet's mass and its distance from the star.

This may seem like a tall order. As we have seen, the light from a planet will be swamped by the light from its star. We can get around this to some degree by using infrared, but even in the infrared the star is millions of times brighter than the planet.

One of the first questions we would have to ask ourselves is, what would we look for? Even if we discovered planets about the size of Earth, very few—in fact, only a tiny fraction—would have advanced forms of life. According to fossil records the first life appeared on Earth about a billion years after it formed, and for the next 2 billion years there was nothing but algae and microbes on the surface. Furthermore, it took another billion and a half years for a civilization to appear. The odds, therefore, if we do detect a planet with life on it, are strongly in favor of the life being primitive. Could we detect it?

Indeed we could. Algae and most microbes add large quantities of oxygen to the atmosphere. They take in water, nitrogen, carbon dioxide, and so on as nutrients and release oxygen. Oxygen is chemically reactive and easily combines with other elements. This means that oxygen in the atmosphere is a clear indication of life. Another sign of life would be the presence of water. Water is essential to life and its presence would be a good sign. Furthermore, we know that plant life uses carbon dioxide in the way we use oxygen, so carbon dioxide would be an indication there was plant life. These three—oxygen, water, and carbon dioxide—are the key. If we detected them on a planet, it would be a strong indication there is life on the planet.

How could we detect oxygen, water, and carbon dioxide? We won't have to image the planets directly; we will merely have to get their spectra. We will have to use the infrared, but this is not a problem; all three of these molecules have strong lines in the infrared. In fact, the infrared spectrum of Earth, Venus, and Mars was taken by the satellite Galileo. Oxygen, water, and carbon dioxide lines showed clearly in the Earth's spectrum, while Venus and Mars showed only carbon dioxide.

One of the major problems in trying to detect oxygen, water, and carbon dioxide in the spectrum of a distant planet is that our atmosphere contains all three of them in excess and would mask anything beyond. We therefore have to get above our atmosphere, preferably some distance out in the solar system. Most astronomers feel that an orbit in the darkest part of the solar system, out beyond Jupiter would be best. The telescope would have to be cooled, but out there it would be naturally cooled.

Even if we did have a telescope in orbit, however, our problems would not be over. If we are to get spectra and direct images of an extrasolar planet we would need a huge mirror, considerably larger than anything that now exists on Earth, and a mirror of this size would be difficult to put in orbit. Roger Angel of the University of Arizona and others have suggested a way around this. As we saw in the last section, interferometry will give the Keck telescopes unprecedented resolution. The same technique can be used in space.

In 1995 Angel suggested that an array of four mirrors would work best, with each of the mirrors a few meters in diameter. They could be arranged in a line—about 50 to 75 meters in length. Several plans are now in the works for such a telescope.

Origins and Horizon 2000

Both NASA and the European Space Agency (ESA) have extensive plans on the drawing board for a search for extrasolar planets and extraterrestrial life. NASA's program is called Origins, and one of its objectives is to put a large interferometer consisting of four telescopes in orbit somewhere beyond Jupiter. It is called *Planet Finder* and is expected to be launched somewhere around 2005.

In 1995 a commission of 135 researchers from 53 universities, federal labs, and companies across the nation drew up plans for what is called the Exploration of Nearby Planetary Systems (ExNPS) program. It is to be a major part of Origins. The objective of ExNPS is to find out how many nearby stars have planets, particularly planets that might be habitable. Of these, the best candidates will be selected for detailed study.

One of the earliest proposed missions will not use interferometry or multiple telescopes. It is a low-cost mission called *Kepler*, but it should be valuable in that it is expected to detect dozens of new candidates. It will be equipped with a one-meter photometer and state-of-the-art imaging equipment. If current plans are carried out, *Kepler* will be put in orbit in 2001.

The ESA's corresponding program is called Horizons 2000. The mission of this program is also to detect terrestrial-like planets in orbit around stars beyond our solar system. The ESA also has plans to use an interferometer. One of the first proposed missions has been named *Darwin*. It will be a large infrared interferometer and is scheduled for launching around 2010.

With these launchings much will no doubt be learned; further extra-solar planets will be found and perhaps the first real evidence of alien life.

chapter 10

SETI: Search for Extraterrestrial Intelligence

We have seen that there is now considerable evidence for planets around nearby stars. And I'm sure if we found a planet that was within the ecosphere of its star and about the size of Earth our first inclination would be to search for signals from it. Fortunately, this won't be a problem as we already have an extensive program of this type in operation.

The reception of a message from space would, without a doubt, be one of the greatest moments for mankind. Even if we had no idea what kind of information was contained in it, it would still be a great discovery, because it would tell us that we are not the only civilization in the universe. Most astronomers are convinced of this now, but the discovery that it is a fact would, without a doubt, have a tremendous impact on us.

What would we do if we received a message? Most people would, of course, want to reply, even though we know that it would take years for the message to get to its destination and a corresponding number of years for a reply to come back. Radio messages travel at the speed of light and the nearest star to us, Alpha Centauri, is 4.3 light-years away. To send a message to a planet orbiting it and receive a reply would take 8.6 years. And it's much more likely that if we did receive a message, it would come from a civilization 20 or 30 light-years away, or perhaps much more. In this case it might take half a century for a reply.

Others say that we shouldn't reply to a message since it would betray our presence to a civilization that is likely considerably more advanced technologically than we are. They fear that such a civilization might attack, enslave, or obliterate us. Movies such as *Independence Day* have indoctrinated us to fears like this. But in reality, unless they knew something we don't about shortcuts through space, parallel universes, and so on, we wouldn't have to worry. Another civilization would be limited to less than the speed of light as we are and therefore wouldn't be able to get here for decades after it received the message, and perhaps much longer. Would it be hostile? We have no way of knowing but can imagine a similar situation here on Earth with the discovery of a lost civilization. At first there would be mostly curiosity; hostility is not likely to come until later.

Project Ozma

The idea of radio contact with an alien civilization has been around for years. Science fiction stories about the possibility were common even early in the century, but strangely the idea was not taken seriously by astronomers. But like many young people Frank Drake was intrigued with the idea. Born in Chicago in 1930, Drake remembered looking up at the stars, sometimes picking out an individual star, and wondering if it harbored

Frank Drake.

life. All through school the question nagged at him. At times he even tried to imagine what aliens might look like, but he never told anyone about his dreams. They might have considered him a little odd.

Drake's interest in science developed early, and was fostered by visits to the Chicago Museum. To his disappointment, however, there were no classes in astronomy at his school, so everything he learned, he had to learn on his own.

When he graduated from high school, Drake didn't think seriously about a career in astronomy. He wanted to become an aeronautical engineer, but when he began applying to colleges, he was disappointed to find that most of them didn't have programs in aeronautical engineering. He thought about his other interests—astronomy and electronics (for years he had loved to tinker with radios)—and finally compromised by signing up for the engineering physics program at Cornell.

He took his first astronomy class at Cornell and was soon hooked. It was during this time that the great astrophysicist Otto Struve came to Cornell for a series of talks. Drake attended them all, and he was impressed. He knew now that he would become an astronomer, but how could he couple it with his knowledge and interest in electronics?

He graduated in 1952 and went into the Navy (the Navy had paid for his education), where he learned a lot more electronics. When he was discharged in 1955 he applied to several graduate schools, selecting Harvard because of its impressive astronomy program.

His adviser at Harvard was the well-known astronomer Bart Bok. Even though radio astronomy was still in its infancy, Bok had started a program in the new area. It had been over 20 years since Karl Jansky detected the first radio signal from space, and radio astronomy had barely gotten off its feet. Grote Reber had built the first parabolic reflector in Wheaton, Illinois, in 1937 and had used it to detect signals from space, but few radio telescopes had been built since.

With his knowledge of electronics, radio astronomy was a natural for Drake. As a new science, it was a wide-open field with unlimited possibilities. But strangely, a number of well-known astronomers had their doubts about it; they were convinced that radio waves would be of limited value in astronomy and that radio astronomy would have a short, sweet life.

Drake graduated from Harvard in 1958 and left for a job at Green Bank, West Virginia, at the National Radio Astronomical Observatory (NRAO). He found the region to his liking; he loved its beauty and isolation. Several telescopes were on the drawing boards, and his job was to set up programs for their use. The search for extraterrestrial civilizations came to mind immediately but he said nothing, knowing that most people would consider it a foolish waste of time. But Drake was patient, sure that an opportunity would eventually come.

Drake had been at Green Bank only a short time when Otto Struve took over as director. He finally got up enough nerve to talk to Struve about his idea, and to his surprise Struve was enthusiastic. He encouraged him to formulate a plan. Drake decide to call the project Ozma, after the queen of the fictional city Oz.

Several telescopes were under construction at Green Bank, but the only one completed was an 85-foot dish. Keeping his plans secret, Drake made all the preparations for the search. Scientists at this time did not talk about the possibility of alien civilizations; it was a topic that most thought belonged in the realm of science fiction. To Drake's delight, however, two well-known scientists, Giuseppe Cocconi and Philip Morrison, published a paper in *Nature* in late 1959 urging astronomers to search. They suggested that radio frequencies would give the best chances of success, and they directed astronomers to a particular region of the radio spectrum.

The day that Drake had been waiting for finally arrived: April 8, 1960. Earlier he had hired two students for the observatory summer program, Ellen Gundermann and Margaret Hurley, and they had volunteered to assist him. He would be using the 85-foot dish and was up at 3:00 A.M., ready to go. His first job was to tune the parametric amplifier, a job he didn't relish because the amplifier sat out from the dish, about five stories above the ground. By 5:00 A.M. everything was ready and Drake and his two assistants were hovering over the chart recorder. They had also set up a loudspeaker so they could hear anything that came in through the telescope.

Two stars had been selected: Tau Ceti and Epsilon Eridani. Both were nearby, only a few light-years away, and both were sunlike. This was important because sunlike stars were the best bets for life. Drake pointed the telescope toward Tau Ceti first; of the two it was the better candidate. The three of them watched the chart recorder in anticipation, expecting something to happen at any moment. But nothing happened and within a few hours their initial enthusiasm began to wear off. By noon Tau Ceti was getting close to the horizon. As it set Drake swung the telescope in the direction of Epsilon Eridani; he barely had it in place when a loud noise came over the loudspeaker. Looking quickly at the chart recorder he saw that it had gone off scale. The two women looked at him, unable to say anything, then suddenly all three of them began to talk at once. Drake was still in shock. Was it really a signal from another civilization? It couldn't be this easy.

One of the first checks in a situation like this is to swing the telescope away from the source and see if the signal goes away, then return it to see if the signal comes back. Drake swung it away and the signal disappeared, but when he swung it back the signal did not reappear.

The team was disappointed and continued to search for the signal over the next few days. In the meantime Drake set up an isotropic (all direction) antenna in the window. If the signal came again and it registered in both receivers, it had to be of terrestrial origin. If it was received only by the dish, it was extraterrestrial. Ten days later the signal reappeared, and to their disappointment it was picked up by both antennas.

Throughout the rest of the project no signals were received, but Drake never thought of the project as a failure. It was a beginning and as such was important, and eventually as news spread, it attracted a lot of attention. The first search for an alien civilization had been made, and it had not been very costly. Ozma was a milestone.

Are There Aliens Out There?

About a year after Project Ozma ended, Drake received a call from Peter Pearman of the Space Science Board of the National Academy of Science. Pearman had heard about Ozma and was intrigued. He had become convinced that there should be some sort of follow-up and suggested that a conference on the search for extraterrestrial intelligence be set up. Drake was receptive to the idea and asked how he could be of assistance. Pearman said he wanted help in making up a list of names of people to invite to the conference—experts in the area. At the time there were, of course, few experts, so the list would not be very long. Pearman also suggested that the conference be held at Green Bank.

Together Drake and Pearman came up with a list of names that included Cocconi and Morrison, Dana Atchley, Barney Oliver of Hewlett-Packard, Carl Sagan, Melvin Calvin, a chemist from the University of California, and a neuroscientist, John Lilly. Drake still had to approach Struve, but to his delight Struve was supportive, so Drake asked him to be host and chairman. With that settled the next problem was the program:

What would they talk about? The major topic, of course, had to be, were there aliens out there, and if so, what were our chances of contacting them? Drake began thinking about it. There had to be a way to determine the probability that an alien civilization existed. He thought about all the factors that would have to go into it. They would be good topics of discussion. Then he realized that these topics were interrelated, each part of an overall probability for the number of civilizations in space. He wrote it down as a formula:

$$N = Rf_p n_e f_l f_i f_c L$$

This eventually became known as the Drake formula and is an integral part of the search for extraterrestrial civilizations today. The symbols have the following meanings:

N = number of detectable civilizations in space
R = rate of star formation in our galaxy
f_p = fraction of them that have planets
n_e = the number hospitable to life
f_l = fraction of these in which life develops
f_i = fraction of these that develop intelligence
f_c = fraction that develop the ability to communicate
L = lifetime of a civilization

The conference was a success, with each of the factors being discussed in turn. Most of the factors are difficult to determine, and the number selected depends on how optimistic or pessimistic you are. The only one relatively easy to determine is R. We know that there are roughly 10^{10} stars in our galaxy and its age is approximately 10^{10} years, so R is approximately 1. Depending on whether you are an optimist or a pessimist, you might select f_p as high as ½ or as low as .005. The same goes for n_e and f_l. Optimists generally select f_i and f_c as 1, so to a first approximation we see that all the factors, aside from L, go to 1. This means from the optimistic point of view N is approximately equal to L, and therefore L is the most critical factor in the formula. Unfortunately, it's also the most difficult to pin down.

How long will a civilization last? The only example we have to go on is our own, and we have had a technology for only a few decades. One of the major factors in determining a civilization's lifetime is its development and use of nuclear weapons. Will a civilization annihilate itself once it has developed nuclear weapons? It looked that way for many years on Earth, but with recent developments things look much more promising. Of course, there are a lot of other problems to consider: energy resources, natural resources, overpopulation, disease, and so on. Most pessimists believe that an advanced civilization is doomed to a few hundred years of life, perhaps even less because of these problems.

Optimists, on the other hand, are convinced that once a civilization gets through a number of early hurdles and manages to live a few thousand years there's nothing to stop it. It will learn to handle problems of overpopulation, annihilation, and energy resources, and will go on to live for millions of years. This means there is a considerable range for L: it could be as low as a few years, perhaps a few decades, or as high as millions of years. And therefore there could be anywhere from a few civilizations out there to millions. On the basis of this it might seem that our formula has been of little help. But as we will see it is still meaningful.

How Should We Look?

As we saw earlier, Drake searched for extraterrestrial civilizations using the large radio telescope at Green Bank. In other words, he used radio waves. Is this the best way? For an answer let's turn to the electromagnetic spectrum. It is the spectrum of photons ranging from short wavelength X rays and gamma rays through to long wavelength radio waves. In between, we have microwaves, infrared, visible, and ultraviolet.

The advantage of electromagnetic waves is obvious. Aside from gravitational waves, they are the only mode of long-distance communication that we know of. And gravitational waves, the waves generated by oscillating masses, are extremely weak, and unlikely to be of any value to us in the near future. One of the major advantages of electromagnetic waves is their speed; they travel at 186,000 miles/sec—the speed of light. With the exception of the hypothetical particles called tachyons, we know of nothing that travels faster. (Tachyons are theoretical particles that presumably

Gamma Rays	X-Rays	Ultra-Violet	Light	Infrared	Micro-waves	Radio Waves

The electromagnetic spectrum.

travel only at velocities greater than that of light, but there is no indication that they exist.)

If we look at the various regions of the electromagnetic spectrum we soon see that the radio region is the one best suited for communication. Visible light is a possibility; we could, for example, communicate with lasers. But lasers have severe restrictions compared to radio waves; they are useless if it is overcast, and laser communication is expensive; furthermore, a laser beam would be difficult to detect (the universe is already full of light beams). Infrared is a region that may become increasingly important in the future, but for now the radio/microwave region is our best bet.

Still, even though this mode appears particularly promising, there are problems. Space is a noisy place. If you hooked the receiver of a radio telescope up to a loudspeaker and scanned it across the frequencies of the radio/microwave region you would hear a continual hiss. And if we are to detect a message from space, we must be able to hear it above this hiss. Listening carefully, however, we soon discover that the hiss is louder in some regions than others. In other words, there is a minumum in the noise (this was pointed out by Cocconi and Morrison in 1959). The region around the wavelength 21 cm (or corresponding frequency, 1420 MHz) is relatively

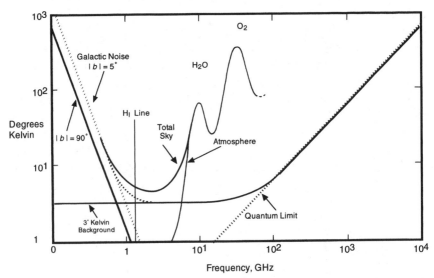

Figure showing the absorption of our atmosphere and from debris in space. The quiet region to the left is called the "water hole."

quiet. Surprisingly, this is where hydrogen, the most common element in the universe, vibrates.

In the low frequency range, below about 100 MHz (MegaHertz, or a million vibrations/sec) our galaxy is a problem, with considerable noise coming from its stars. On the high-frequency end there is absorption by the Earth's atmosphere, and noise from electrons in interstellar gas clouds. And finally the cosmic background radiation causes a low noise across most of the region.

Close to the hydrogen line at 1420 MHz is the OH line at 1665 MHz. The region between these two lines was referred to as the "water hole" by Cocconi and Morrison (H + OH = H_2O), and to them it appeared to be the best place to search. In fact, the water hole seemed to be an appropriate place for alien civilizations to meet, just as early tribes on Earth gathered at water holes. Most of the searches that have been made, have been in this region.

What Should We Look For?

Why should we expect space to be full of messages sent out by alien civilizations? We have the capability of sending out messages, yet we have sent very few. It is possible that everyone is like us, each listening for messages sent by others, with no one sending messages.

Assuming this isn't the case, what type of message would we expect? It's easy to see that there are three possibilities. Messages may be purposely beamed into space by advanced civilizations. We can't be sure why they are being sent, but if this is the case, the civilization sending them is likely more advanced technologically than us. Nikolai Kardashev of the former Soviet Union classified extraterrestrial civilizations into three types, according to their energy resources. Type I civilizations would be more or less equal to us technologically. Type II would control and utilize the entire power output of their home star. And Type III would control the power output of all the stars in their galaxy. Type II are of particular interest to us because they are the most likely to be beaming a signal into space. Furthermore, they are most likely to be associated with the second of our three possibilities, the possibility that some civilizations are communicating with one another regularly across the galaxy. Type II and Type III civilizations, and perhaps a few Type Is may be doing this, and if so, we might be able to intercept one of their messages.

Has there been enough time for Type II civilizations to have arisen? Indeed there has. Our galaxy is approximately 15 billion years old, and if life on Earth is representative, it takes about 5 billion years for an advanced life form to arise (assuming all the conditions are right). We have to assume we are late bloomers, since we have had an advanced technology for only a few decades. What about the first 10 billion years our galaxy existed? Many civilizations should have arisen during this time if it only takes 5 billion years to produce intelligence, and some of these civilizations would now be 2 billion or 3 billion, and maybe even 5 billion years old. A civilization this old would certainly be Type II or III.

Interception of a message between two civilizations is obviously one way of locating extraterrestrials, but surprisingly there is another, even better method. Not all signals in space are purposely sent; there is a lot of "stray" radiation. The presence of this stray radiation was first brought to our attention in 1978 by W. T. Sullivan of the University of Washington and two of his students. Sullivan's paper, which appeared in *Science*, addressed the question: How much radiation is leaking from Earth, and what is the possibility that other planets have a similar leakage? As Sullivan pointed out, some of the radiation from television stations is picked up by televisions on Earth, but much of it leaks off to space. In fact, all short wavelengths penetrate our atmosphere, and therefore CB radio, FM radio, pulsed military radar, as well as television transmission, leak to space.

The most powerful source of radiation that is leaking is military radar, but it is pulsed and therefore not likely to be interpreted as an alien signal. A television signal, on the other hand, could easily be recognized as arising from a technology. The first high-frequency radio stations came on-line in the 1930s and the first television stations came little more than a decade later. This means that the Earth is now surrounded by a rapidly expanding radiation bubble approximately 60 light-years in radius, and any civilization with a large radio telescope within this bubble could detect it. On the other hand, if a nearby star had a planet with an advanced civilization, it too would have an expanding bubble around it, and when the outer limits of this bubble passed us, we would be able to detect it.

What does the Earth's bubble look like? It arises from thousands of television stations on earth, each sending out a signal of relatively narrow bandwidth. The major concentrations are in the eastern United States and Europe, and as the Earth spins on its axis a distant civilization would see two "hot spots." With an average-sized antenna, a nearby civilization could easily pick up this radiation, but with so many signals, all superimposed, it would be a general "mush" to that civilization.

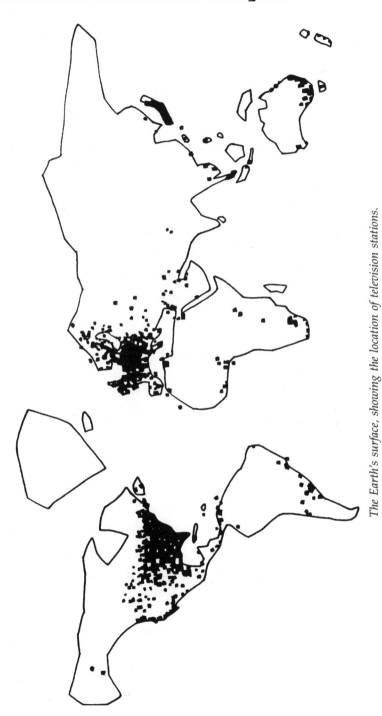

The Earth's surface, showing the location of television stations.

An interesting question is, could the civilization tune into individual stations and receive some of our early programs such as "I Love Lucy" and "The Beverly Hillbillies?" This is possible with a large enough array, and we might expect Type II civilizations to have arrays of this size, but in general it would be difficult. A star about 25 light-years away with a large array could, however, learn a lot about us even if the civilization couldn't tune into individual signals. They could learn that the Earth spins in 24 hours, and if they watched us over a few decades, they would see that our technology was advancing rapidly. Between 1940 and 1970, for example, the "brightness" of the Earth increased by a factor of 1000. They would quickly determine that the signal was made up of many different frequencies, as each station broadcasts on a different frequency. This would suggest that the signal came from many different sources.

The signal would also vary because of the Doppler effect. As we saw in chapter 9, the Doppler effect is the shift in frequency in a signal that occurs when a source is either approaching or receding. You are likely familiar with the effect in connection with the change in pitch that occurs in a car horn as it approaches or recedes from you. The two "bright spots" on the Earth would be approaching the planet during part of the day, and receding from it during the other part because of its spin. In addition, the Earth also goes around the sun, and this would also cause a Doppler shift. A civilization on a planet about 25 light-years away could easily measure both of these. Thus, in addition to determining the length of our day, and our year, it could determine the size of Earth, and that we were in the sun's ecosphere. Turning things around, we could do the same if we located a civilization about 25 light-years away.

Stray radiation from a planet is probably the best thing to search for, because it is the most likely thing we will find. Unfortunately, as a civilization becomes more advanced it will probably stop any radiation from leaking to space in an effort to conserve energy.

What Would We Ask?

Let's assume that we got a signal from space and determined that there was a civilization about 15 light-years away. It would likely be more advanced technologically than us, and there would be a lot we could learn from it. It would, of course, take years to get a reply to a question, but for now let's not worry about this. Let's consider what sort of questions we would ask them.

We could begin by asking them about their chemical basis and technological level. Also of considerable importance to us would be some of the central problems in science. We might ask, for example, about their basic scientific theories. Do they have a theory similar to our general relativity and quantum mechanics? Have they developed a unified theory—a theory of everything—and if so, what form does it take? We could ask if time travel is possible, and inquire about their understanding of the origin and structure of the universe.

Other questions of tremendous importance to us are these: What sources of energy do they use? How did their civilization last so long? How have they overcome problems of overpopulation and so on? There are also critical questions in relation to disease. We could ask them if they had cured cancer, and other diseases. We could also ask them about the function of the brain.

Many of these questions would make little sense to them. Their brains are likely quite different from ours; they may also have different diseases and their science may be quite different. In addition, it would be disappointing to many to have solutions to age-old problems handed to us. We would lose the thrill of discovery. On the other hand, some of the answers might be irrelevant by the time we got them. If the civilization was 15 light-years away, it would be 30 years before we got an answer, and we may have discovered the answer ourselves during this time.

The biggest problem, however, would likely be understanding the replies. Maybe they are too far beyond us or too "different" for us to understand what they are talking about. It might be like trying to explain "string theory" to a native of a forgotten tribe in the interior of Borneo. String theory is our current best candidate for a "Theory of Everything."

Interpreting and Decoding the Message

Let's assume we received a message that was purposely sent. There's little doubt that it would be incomprehensible at first. Aliens are bound to be quite different from us; in fact, we have little idea what they would be like, so we can only guess at what they would include in their message.

Decoding, or making sense out of the message, would not be easy. When we send a message to a distant point on Earth we superimpose it on an electromagnetic wave by modulating the frequency. But the person who receives the message knows what frequency we are sending and knows the bandwidth and how the message is modulated, so it is no problem for that person to receive and decode it. This is not possible across space, between two entirely different cultures with different languages.

With a little perseverence we may eventually be able to decode the message, or at least part of it. We have had some experience in this. During World War II, for example, the Allies were relatively successful in decoding secret German and Japanese messages. But we haven't been very successful with early civilizations; many had languages that we still don't understand. And our partial successes have, in many cases, been based on luck.

How would we go about trying to decode a message that came from outer space? If the message was purposely directed, we could assume that the civilization made it as simple as possible. On the other hand, if we were eavesdropping on messages sent between two civilizations, or detected their stray radiation, decoding would be much more difficult. Fortunately, there are things that are common to all civilizations. Mathematics is universal; the constants of the universe (things such as the speed of light, the charge on the electron, or Planck's constant) would have to be known by them. Some of the basics of science are also universal: the 21 cm line of hydrogen, the elements, and some of the laws of nature. If they wanted us to understand the message, some of the above would no doubt be included in it.

If the message was purposely beamed it might be in the form of a "call signal." This is a simple message that is beamed over and over, and in most cases directs the receiver to a more complete message at a different frequency. The technique is commonly used on Earth.

If we received a message, one of the first things we would have to do is "clean" it up. All signals are distorted to some degree as they pass through gas clouds and so on in their trip through space. Also, it would have to be separated out from the background noise.

One way of determining what might be in the message is to consider what we would send, and as it turns out we have sent several messages to space.

Sending Messages to Space

One of the first to think about sending a message to space was Frank Drake. He experimented with black and white images, binaries that could be transmitted using zeros and ones. He sketched out a pictogram using the idea. It included a schematic of the solar system, a crude diagram of the oxygen and carbon atoms, a representation of the numbers one to five in binary form, a representation of human figures, and other numbers. Then he sent it out to several fellow scientists to see if they could decode it. Almost no one decoded it completely.

Barney Oliver improved on Drake's pictogram, keeping most of what Drake had incorporated, but he showed two humans, a male and a female, with a child between them. It was much easier for most people to interpret. It was intended as a message to be sent into space using a radio telescope, but interestingly the first messages sent to space intentionally were not radio messages. They were plaques aboard *Pioneer 10* and *11*. Upon hearing that *Pioneer 10*, which was launched in 1972, would eventually leave the solar system, Carl Sagan realized that one day in the distant future it might crash into a distant planet or be intercepted by a civilization. He convinced NASA to include a plaque on it with a message to the civilization that encountered it. Some of the things that were on the pictogram were included—a representation of the solar system, and a male and female figure. Interestingly, the two naked figures created quite a stir within a small sector of the general public (they were published in a number of major newspapers). I wonder if it really would have helped if we had clothed them. Anyway, a similar plaque was put on *Pioneer 11*.

The first radio message was sent to space in 1974. Over a period of three years the 1000-foot Arecibo radio telescope in Puerto Rico had been taken apart and upgraded. When completed, it was a new, much more powerful instrument and officials decided to rededicate it. Drake, now director of the observatory and responsible for the ceremony, decided that it would be appropriate to beam a message to space. He took another look at the pictogram he had devised earlier, and at the one devised by Barney Oliver, and decided that a slight modification of them would be appropriate. This one showed the basic molecule of life on Earth, DNA, along with human figures, the solar system, and a simple representation of a telescope.

A target was selected: the globular cluster M13 in the constellation Hercules. It is at a distance of 24,000 light-years so it would be a while before the message got to it, but it was considered to be an appropriate target because of the large number of stars (a few hundred thousand) that would be included within the beam. A loudspeaker was set up so that the 250 guests could hear the message as it was transmitted. It took three minutes to send.

Finally, an elaborate audiovisual disk was put aboard *Voyager 1* and *2*. It included many images from Earth: flowers, trees, animals, oceans, deserts, houses, and so on. And it included many sounds such as laughter,

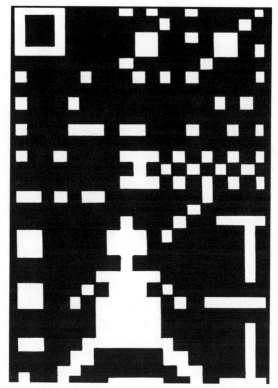

Pictogram of the type sent by the Arecibo telescope.

thunder, the chirping of frogs, several musical selections, along with short greetings.

SETI Programs

As we saw earlier, Project Ozma got a lot of attention, both in the United States and abroad. But strangely it was many years before another search was made in the United States. The major Search for Extraterrestrial Intelligence (SETI) programs over the next few years were in the U.S.S.R. Two names were prominent in the early Russian program: Kardeshev and Troitsky. Nikolai Kardeshev, as we saw earlier, was the originator of Type I, II, and III civilizations. He was involved in a project in 1964–1965 in

The Arecibo radio telescope. (Courtesy National Astronomy and Ionosphere Center)

which two distant quasars were searched for signals. For a while the team believed that they had received a signal. Vasevolod Troitsky was involved in several projects during the 1960s. In most cases the Soviets hunted for pulsed signals, but none were found.

The follow-up of Ozma in the United States came in 1971–1972 when Gerritt Verschuur at NRAO searched the vicinity of nine stars. Verschuur called his project Ozpa (a pun on Ozma). He used the 140-foot and 300-foot telescopes at Green Bank. A much more extensive search was initiated in 1972; it lasted until 1976. Ben Zuckerman of the University of Maryland and Patrick Palmer of the University of Chicago used the 300-foot telescope to look at 674 stars near the 21-cm frequency. Again no signals were found.

About this time John Kraus and Robert Dixon started the Ohio State All Sky Survey. This program is now the longest running, continuous program in the United States; it is still active. Their radio telescope, called "Big Ear," is different from most; it is a huge antenna with a receiver that covers approximately 20 acres, and it is rectangular rather than round, as most other radio telescopes are. The intriguing "Wow" signal was received here. Jerry Ehman scribbled "Wow" next to it when he discovered it in 1977. It was one of the highest-intensity signals ever detected by the receiver and has never been detected since.

One of the major difficulties in searching for a signal from an extraterrestrial civilization is the number of frequency channels that must be examined. We can only guess at the bandwidth of a signal, but most natural signals are broadband, so it seems likely if a civilization was purposely beaming a message to space they would restrict it to a narrow band of frequencies. A narrow band would be much easier to distinguish from natural signals. Although this appears to be a reasonable conclusion, it creates a severe problem for us, since the narrower the frequency band, the greater the number of bandwidths that have to be searched. Many astronomers believe that a bandwidth of 1 Hz would be typical of what would be sent. There are, unfortunately, 10 billion channels of this size in the microwave region alone, and it would take years to search each one properly. And this would be for only one star, or one tiny region of the sky. We would like to survey many stars, and eventually do an entire sky survey. It goes without saying that the task would be overwhelming.

Fortunately, there is a way around part of this problem. Many different frequency channels can be searched at the same time using a device called a multichannel spectrum analyzer. Over the years there has been a

Paul Horowitz.

steady increase in the number of frequency channels that can be searched simultaneously. Today it is in the billions.

One of the brightest stars in multichannel analysis is Paul Horowitz of Harvard. He has spent most of his career improving multichannel analyzers. He began his work in 1981 when he received a NASA/Ames fellowship that allowed him to work on a SETI program at Ames Research Center and Stanford University. He was part of a group that upgraded an existing analyzer to 128,000 channels.

Horowitz was able to make this device so compact that he could fit it in a suitcase. He dubbed it "suitcase SETI" and took it to Arecibo in 1982 where he examined 250 stars using it. None gave any indication of signals. In 1983 he took it back to Harvard where he set up a permanent program using an 84-foot radio telescope. It became known as Project Sentinel.

During the early years of SETI many searches were also made outside the United States, in France, the Netherlands, Germany, and Canada. No alien messages were detected in any of the searches.

Cyclops

Although it is now only a past unfulfilled dream, the gigantic project called Cyclops is worth mentioning. In 1971 a panel of NASA scientists looked into what was possible in the way of a large telescope using available technology. The project they came up with was Cyclops. It was to be built in stages over several years and would have dwarfed anything else in the world. It was to consist of 1500 antennae, each 330 feet (100 meters) in diameter; the overall array would have a diameter of about five miles and would be equivalent to a single dish of that size. The cost would have been as great as the cost of the Apollo project—about $20 billion then (considerably more in today's dollars).

Cyclops would have had tremendous potential, allowing us to contact and communicate with civilizations hundreds of light-years away. But the price tag proved to be too great, and although many scientists pushed for it, it was never funded.

One of the difficulties was the grandiose scope of the project. It was to be built gradually, starting with only a few dishes, with a few more being added each year. But most people in Congress saw only the entire project—all 1500 dishes—and the enormous price tag.

SERENDIP, META, and BETA

In the mid-1970s Jill Tarter and Stuart Bowyer of the University of California built a device that could ride piggyback on a radio telescope without interfering with the ongoing program. They first used their device on the 85-foot radio telescope at the University of California's Hat Creek Radio Observatory. Attaching the device to the telescope, they received the same signal as the primary researchers, then analyzed the signal for indications it was from an extraterrestrial civilization. The project was called Search for Extraterrestrial Radio Emissions from Newly Developed Intelligent Populations (SERENDIP). This was a new and economical way of searching that didn't tie up a radio telescope. In fact, it was also made completely automatic, so the researchers could work on other projects while it was in operation. Tarter was, in fact, involved in another search with several colleagues in which 200 stars were searched while SERENDIP was in progress.

In 1985 SERENDIP moved to Green Bank where it was attached to the 300-foot telescope. Here it became SERENDIP II. But the ungainly, and poorly constructed, 300-foot telescope collapsed in 1989, so SERENDIP had to move again. This time it went to Arecibo where it became SEREN-DIP III. It logged 10,000 hours on the huge Arecibo telescope and checked the entire visible sky from Arecibo at least once (some parts twice), but it found nothing. SERENDIP IV is now in progress at Arecibo.

After Paul Horowitz set up Project Sentinel at Harvard he continued working on his multichannel analyzer. (Project Sentinel, incidentally, was funded by the Planetary Society, a society that was set up for research in the solar system and to help SETI, by Carl Sagan and Bruce Murray.) Horowitz's aim was to increase the capability of his analyzer to 8 million channels.

META, as the new analyzer was called, was completed in 1985. (It was paid for by a donation from the movie director Steven Spielberg.) It was used for an all-northern-sky search between 1985 and 1994. In the analysis of five years of data, during which 60 trillion channels were searched, 37 interesting signals were found. None of them have been repeated, however, and there is nothing definite to indicate that any of them is from another civilization.

While this was going on, Horowitz was already thinking about extending his analyzer to a billion channels, and he soon began working on it. In 1995 he achieved his goal as BETA was unveiled. It took four years to design and build, with most of the funds coming from private sources.

NASA-SETI: MOP

Despite the failure of Cyclops and funding difficulties with other projects, scientists at NASA continued their interest in SETi, and by the early 1990s a new project was in the making. Called HRMS (High Resolution Microwave Survey) when it was initiated in October 1992, its name was changed to MOP (Microwave Observing Project) soon after; funding of $10 million a year was allotted for it. It was to be a two-phase program. Half of it was to be a targeted search, centered at Ames Research Center. About 800 stars within 75 light-years of the sun were to be searched using some of the world's largest radio telescopes: the 1000-foot dish at Arecibo, the 140-foot dish at Green Bank, and the 210-foot dish at Parkes, Australia. The second part of the program was an all-sky search. It was to be centered at JPL and would have used NASA's deep space network satellite tracing antennas. A multimillion channel analyzer was to be used in the project.

Within a year of its initiation, funds for MOP were withdrawn. Because of this, the SETI Institute was set up. Using private funds the SETI Institute at Mountain View, California, has gone forward with the targeted search. Current plans call for the search of about 1000 nearby sunlike stars. The new project is called Phoenix. Several telescopes will be used in this project, including the Parkes and Mopra telescopes in Australia, the 140-foot NRAO telescope, and the Arecibo telescope in Puerto Rico. The main radio telescope involved in the search is the 140-foot telescope at Green Bank, West Virginia. Several candidate signals per week are detected by this system, but so far none has panned out as a genuine signal from an ETI.

In conclusion, it is important to mention that the passage of time is perilous to SETI. With the advancement of technology, the space around us is rapidly getting noisier. Within a few years it will not be possible to make SETI searches from earth. We will then have to go to the far side of the moon—a much more expensive endeavor than MOP, or any other Earth-based program.

chapter 11

Are We Alone?

Nobel laureate Enrico Fermi and several colleagues were discussing the latest scientific problems over lunch one day in 1950 when the topic "extraterrestrial life" came up. Was there any evidence for intelligent life beyond Earth? Could we say anything about the possibility? As was typical of him, Fermi made a few quick calculations in his head related to the number of stars in our galaxy, its age and time it takes for life to evolve, and it soon became clear to him that something was wrong. If there were other civilizations in our galaxy, many of them would have to be much more advanced technologically than we are. After all, our galaxy is approximately 15 billion years old, and the Earth has only been around for 4.6 billion years. Fermi became convinced that a "wave of colonization" should have swept through our galaxy by now. "Where are they?" he said jokingly to his colleagues. "They should have been here by now." These three seemingly innocent words, "Where are they?" is now his most famous quote.

As we will see there is, indeed, a problem, and many people have begun to consider it. We frequently think of the number of stars in our galaxy—200 billion—as so great that civilizations could easily get lost among them, and even if some of them did move out to nearby stars and colonize them, their little corner of civilization would be nothing more than a tiny speck in our galaxy, so small it would be difficult to see. But a few simple calculations show that this may not be the case.

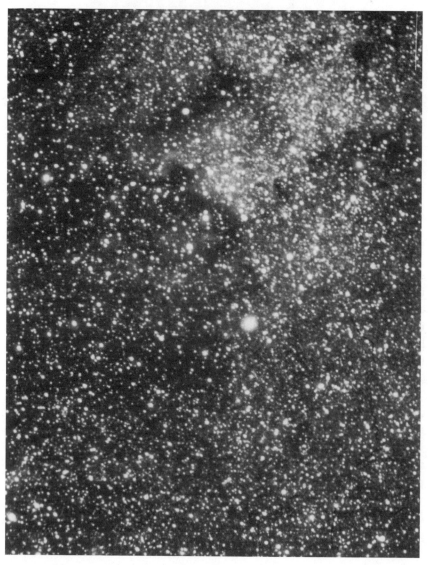

With the large number of stars in this photo it would seem unlikely that we are alone.

Tipler's Argument

There are several ways of showing that Fermi's question is valid and that it has important implications for the SETI program. Frank Tipler of Tulane University was one of the first to follow Fermi's lead. He was disturbed by the large number of extraterrestrial civilizations Drake and Sagan were predicting. Using Drake's formula, they said that there should be about a million civilizations in our galaxy. To Tipler this number was outrageous.

In several articles beginning about 1980, Tipler went to the opposite extreme and predicted that our civilization was the only one in our galaxy. To many people this seemed crazy, but Tipler was convinced that he was right. He based his conclusion on two things: the small probability that advanced forms of life will evolve, and the high probability that if they did arise, they would spread out and colonize the entire galaxy.

Tipler pointed out that we have the capability of sending probes to the stars, and indeed had sent two spacecraft (*Pioneer*) to deep space. With their slow speeds it will take a long time before they reach the nearest stars—about 80,000 years—but compared with the age of our galaxy this is a relatively short period of time. Of more importance, as far as Tipler was concerned, is that in the near future we will be capable of sending probes into space at much greater speeds. Speeds up to 10 percent that of light will be reached during the next century, and they will allow us to reach nearby stars in 50 years.

If the trip to a nearby star took only 50 years it seemed reasonable to Tipler that a wave of colonization would spread out rapidly from any advanced civilization. He acknowledged that few people would volunteer for a flight that would take most of their lives, but he pointed out that there was a way around this. Our computer technology is advancing rapidly, he said, and within the next century we will likely have computers that can reproduce themselves. John Von Neumann predicted such devices many years ago. Indeed, according to Tipler, if a computer could reproduce itself, it could do many other seemingly miraculous things: it could be programmed to build a smelter from raw materials, and to build a spaceship from the resulting metal, and to guide this spaceship through space to a specific destination.

Tipler admitted that advanced civilizations may be reluctant to expand to the stars with humanlike volunteers, but they could easily send spaceships controlled by robots to them. According to his plan, when these robots got to another star they would land on an orbiting planet, colonize it with robots, and make new spaceships. At each landing several new missions would be sent out, and when they landed they would continue the process. This would give rise to what is call an "exponential increase" in the number of missions, and when things increase exponentially, they become large rapidly. A good example of an exponential increase is starting with a penny and doubling it each day (1, 2, 4, 8, 16, 32,...). If you made this much money you would be a millionaire by the end of the month. Similarly, with 30 turnovers in Tipler's scheme, millions of probes would be sent out, and if you assume each turnover took only a few hundred years, our entire galaxy would be colonized in a few million years, which is far short of the age of the galaxy.

But we see no evidence of this "wave of colonization," so Tipler asks: Why hasn't it reached us? His answer is that we don't see it because it hasn't occurred, and this of course implies that we are the only advanced civilization in our galaxy.

It's relatively easy to poke holes in Tipler's idea. One problem we see immediately is that many of the colonists would not send out probes, just as in a chain letter scheme many people do not bother to continue the chain. But with the time we have to work with, this would make little difference, according to Tipler. If only a small fraction continued it, our galaxy would still be colonized in little more than 100 million years.

A variation on Tipler's scheme—if you don't have faith in computer robots—would be to carry fertilized eggs, then incubate them when you get to your destination. In this way you could colonize the planet with people.

Papagiannis's Argument

Tipler's argument may be difficult to accept because it is based on computers that have the intelligence and abilities of humans. To many it might seem unlikely that this will ever come to pass. Michael Papagiannis of the University of Boston gets around this, however, and still comes to the same conclusion.

As we will see in chapter 12, Gerard O'Neill of Princeton University outlined a plan in the 1960s for putting large space stations in orbit, stations capable of housing and sustaining 100 to 1000 people in an Earthlike environment. A hundred years from now, according to Papagiannis, such colonies will be common. And as these space colonies grow larger and become more sophisticated, they will eventually move from their Earth orbits, taking up residence around the sun. Gradually they will move to the edge of the solar system, and finally to interstellar space. With the use of nuclear fusion, it will, in fact, be relatively easy for space colonies to become interstellar sojourners. Their only need would be an occasional visit to another solar system to get raw materials and fuel.

Whole generations will pass their lives in these large vehicles, never knowing anything but life in them, so there will be no problem with volunteers. If they travel at, say, 2 percent the velocity of light, they will get to nearby stars in 500 years. If we assume they set up a colony and spent 500 years establishing it before sending out other probes, it is easy to show that a wave of colonization would sweep across our galaxy at a rate of about one light-year per century. This may seem slow, but it is all that is needed to colonize our entire galaxy in 10 million years. This means intelligent beings on every habitable planet in the galaxy! Because of their connection, such beings would be aware of other alien colonizers and would no doubt be in communication with them, so there would be a "galactic club" throughout the galaxy.

Again, we can imagine that many of the colonizers would break the chain, but to stop the entire wave of colonization everyone would have to be stopped, and with our natural inclination to explore, this seems unlikely. Colonization of our entire galaxy would still occur if only a few percent continued sending out probes.

Papagiannis also points out something else that is frequently overlooked. If there are 1 million advanced civilizations in our galaxy now, as Drake and Sagan have predicted, then there would have been a billion such civilizations in the past. (Drake's equation gives the number of civilizations we could communicate with *now*.) Furthermore, it would take only one of these billions to start a wave of colonization, and the galaxy would be full. It's easy to see from this that there is a problem. Either colonization has taken place and there are civilizations all around us, and for some reason or other we haven't detected them, or no colonization has taken place and we are the only advanced civilization in our galaxy.

In theory, then, we shouldn't have to search thousands of stars over decades to find out if there are extraterrestrial civilizations among them. We need look at only a few of the nearby stars. If there is no indication of life among them, it's unlikely our galaxy is colonized, and we are alone.

It might seem that this is pushing things too far. But consider the Earth. We have made tremendous technological advances over the past 100 years. Before that, in fact, we had virtually no technology. Similar advances likely take place on other planets assuming their civilization is similar to ours. And, as I emphasized earlier, 100 years is barely the blink of an eye on the cosmic time scale.

Criticisms

Both Tipler's and Papagiannis's arguments, as you might expect, have drawn criticism. Drake has pointed out problems in Tipler's program. The major difficulty, as far as he is concerned, is the spaceships. Drake concedes that computers may eventually be able to reproduce themselves. But for colonization, much more than this is required. The robot must decide on a suitable planet within a given system, land on it, then using the available raw material, set up a colony, a smelting plant, and a factory for making other robots and spaceships. As Drake points out, most of the problems that would be encountered are not even predictable, and yet these robots must reason and make decisions the way humans do.

Another difficulty is the acceleration and deceleration of the spaceships. According to Drake's estimates, the energy required for such a flight would be equal to the total energy output of the United States over a period of 100 years. He doubts that any civilization would agree to expend such a large amount of energy on a project such as this.

Both Drake and Gregory Benford of the University of California at Irvine point out that the motivation for such projects may be low. These voyages would take almost 1000 years, and the real payoff may not come for 10,000 years. Tipler states that such a program would cost $30 billion—about the cost of the Apollo program. But in the Apollo program we saw results almost immediately. Would a civilization put out $30 billion if the results came in 1000 years? It seems unlikely.

Furthermore, over such a long period of time the probability that the computer would malfunction is high. Computers don't operate forever here on Earth, and it's unlikely they will in space.

Papagiannis's program is, perhaps, a little easier to swallow than Tipler's because it doesn't require supercomputers with the intelligence and reasoning power of humans, but difficulties also exist in it.

Another Look at Drake's Equation

We saw earlier that from an optimistic point of view Drake's equation predicts up to a million civilizations in our galaxy that we could communicate with. If this is true, the nearest civilization is, on the average, about 100 light-years away, and we should be able to detect it. But let's take a closer look at Drake's equation. We have, in reality, glossed over a number of thorny issues.

We can divide the factors in the equation into two types: those based on astronomy and those based on biology. The factors based on astronomy are R (the rate of star production in our galaxy), f_p (the fraction of stars that form planets), and n_e (the number of these planets in the life zone of the star). In most cases R is given a value 1, and f_p and n_e are given values slightly less than one. As a group they may give a factor of, say, $1/100$. This may be slightly in error, but it's not dramatically off since it's based on numbers that are relatively secure. Our real problem is the next two factors (we'll neglect L for now): f_l (the fraction of planets on which life forms) and f_i (the fraction of these that develop intelligence). Both of these numbers are based entirely on biology, and are therefore much more difficult to determine accurately. Earlier, from an optimistic point of view, we gave each of them a value 1. Biologists, however, are far from convinced that this is valid. And, indeed, there are a lot more uncertainties in these numbers than there are in the previous ones.

Let's consider f_l first. This is the probability that life will arise on a suitable planet in the star's ecosphere. A lot of factors go into the determination of this number. Earlier we assumed the ecosphere of a star like our sun is relatively wide, extending roughly from Venus's orbit to that of Mars. Calculations by Michael Hart of Goddard Space Flight Center in 1978 indicate, however, that the sun's ecosphere is actually much narrower than this. Hart showed that if we were 5 percent closer to the sun or 1 percent farther away from it, conditions would not be satisfactory for life. More recent calculations made by Jim Kastings and several collaborators show, however, that the ecosphere is slightly wider than Hart estimated. Still, many biologists are now arguing that natural evolutionary generation of life (the process that occurred on early Earth) may be a much more improbable event than astronomers believe. With the narrow exosphere and other factors taken into consideration, biologists feel that a more realistic estimate for f_l is $1/1000$.

The factor f_i is, in many ways, even more difficult to determine than f_l. It is the probability that unicellular organisms eventually give multicellular forms, which in time produce intelligence. On Earth this took 3 billion years, and it required a lot of special conditions. The temperature had to remain relatively uniform over this time, the Earth's atmosphere had to change from reducing to oxidizing, and an ozone layer had to form to protect life from the UV radiation of the sun. The probability of all these things happening, at the right time, in the right way, is obviously very small. Few biologists would give it a value 1; most believe it is also closer to $\frac{1}{1000}$.

Multiplying these two factors (f_l and f_i) together, we get a millionth the number of civilizations that we got previously. That leaves us with one—namely, us. This is another argument that we may be alone in our galaxy. But again there are many uncertainties, and biologists may be estimating much too low.

Longevity

In addition to the uncertainties in f_l and f_i, one of the other factors in Drake's equation, namely, L, the estimated lifetime of a civilization, has considerable uncertainty. In arriving at a million civilizations in space, we assumed that a typical lifetime was a million years. Are we being too optimistic?

The only example of a civilization we have is our own, so let's consider it. We have had a technology for perhaps 100 years. How much longer will it last? To answer this, several things must be considered, with population control near the top of the list.

Overpopulation is one of the most serious short-term threats to Earth. As strange as it may seem, our planet is almost full now, in the sense that it is not able to support many more people. Earth's population doubled every 40 years until about 1950. Since then it has increased its doubling rate to about 35 years (in some countries it's as low as 20 years). Looking at a plot of the Earth's population (see the figure) it's easy to see that the increase is dramatic. In fact, extrapolating the graph we see that the population of the Earth will reach infinity in the year 2030. (This, of course, won't happen; something will have to give.)

I was reading Robert Fulghum's book *All I Really Need to Know I Learned in Kindergarten* a few years ago when I came across the prediction that at the present rate of expansion the total mass of human flesh and blood will be equal to the mass of the universe in the year 6826. At the time I chuckled, but it's easy to show that it is roughly correct, and 6826 is not that far in the future.

Population growth is obviously a serious problem, and coupling it with the fact that the Earth has finite resources, it becomes even more

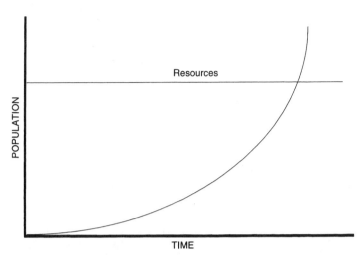

A plot of population versus time with resources shown.

serious. A catastrophe of some sort will obviously occur when the population growth passes the line representing our resources (see the graph). Computer models have been made to see what will happen if the present trend continues unabated. Before I get to it, I should mention that there are other factors that are important, one of the most serious of which is pollution. As the population increases, pollution of our air and water will also increase, and it will have a serious impact on population growth.

From the graph (p. 194) we see that somewhere between 2000 and 2100 the Earth's population will fall catastrophically to a small fraction of what it was earlier. By careful planning—watching population growth, controlling pollution, increasing food production efficiency, and recycling—we can change this significantly; indeed, we can even level off the population curve. There is little indication so far, however, that we're doing much to achieve this. Furthermore, although we may be able to extend the inevitable for a century of two, we are still a finite system, with finite resources, and sooner or later they're going to run out.

The ideal would, of course, be a steady state situation, in which each birth is balanced by a death. With this, and the use of sunlight as an energy source, we could survive far into the future (to do this we would eventually have to trap all the light from the sun). And of course we have the further option of expanding out into the solar system and using the resources of other planets, moons, and asteroids.

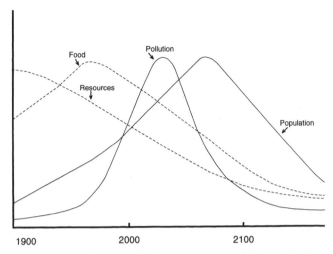

Plot showing population, pollution, resources, and food in the future.

To get a million civilizations in our galaxy we needed L to be a million years. But as we have seen, we are having trouble stretching the Earth's future more than a few centuries. Is it possible that the Earth could survive for a million years? It seems unlikely, but we can't say it's impossible. It would likely require expansion to the planets of nearby stars. There are a lot of uncertainties. Based on what we have seen, however, 10,000 or perhaps 100,000 years appears to be a more realistic number for L. This would give 10,000 to 100,000 civilizations (assuming optimistic values for other factors).

The Anthropic Principle

Tipler gives another argument for a small number of civilizations in our galaxy. It is based on the anthropic principle, which was developed by Brandon Carter of Cambridge, John Wheeler of Princeton, and others.

To understand the anthropic principle we must begin with the Big Bang theory. According to this theory, the universe began as a huge explosion, an explosion that is still going on, with all the galaxies expanding away from one another. One of the fundamental questions of cosmology is this: Will the universe continue to expand forever, or will it eventually collapse back on itself? This, it turns out, depends on the average density of matter in the universe: if it is greater than a certain critical value the universe will collapse back on itself, if not, it will expand forever.

John Wheeler and others have shown that if the universe collapses back to a singularity (a point of infinite density), the fundamental constants of nature (things such as speed of light, mass, and charge of the electron, and Planck's constant) would be lost, and if the universe reemerged from the singularity, it would reemerge with different fundamental constants. Carter looked into the effect of new fundamental constants on the distribution of stars in the universe. He found that if the constants were adjusted slightly in one direction, certain types of stars could not exist (say, red dwarfs). And if they were adjusted slightly in the opposite direction blue giant stars couldn't exist. But blue giants are critical to the production of heavy elements, and without these elements, life would be impossible. The universe, and the life within it are, apparently, strongly tuned to the fundamental constants it has. And since different fundamental constants mean a different-sized universe, life as we know it is also tuned to the size of our universe. According to Tipler, the universe must contain 10^{20} stars in order to contain a single intelligent species. If so, it is not surprising that we are the only civilization in our galaxy.

Drake does not agree with Tipler's conclusion. He points out that we still have only a partial understanding of the anthropic principle and its region of validity. He concedes that the idea is interesting, but he finds it difficult to believe it has a strong bearing on the number of civilizations in the universe.

Neo-Darwin Contingency

Biologists have another argument that limits the number of extraterrestrial civilizations. It is referred to as the neo-Darwin contingency, and as the name suggests it is related to Darwin's theory of evolution. Most biologists are convinced that evolution is a random process. In other words, the changes that occur are entirely random. If this is, indeed, true, then the long evolution from unicellular species to complex, organized, multicellular species has to be an accident of nature and will never happen again. In short, it is such a long shot, it could never occur twice. This would, of course, reduce f_i dramatically, and the probability that there is another advanced civilization in space would be close to zero.

But we know that evolution has provided particularly complex life forms here on Earth. So perhaps evolution is not as haphazard as biologists believe, perhaps it does change along certain paths, many of which lead to intelligence. Something may be built into the laws of nature to give this. Indeed, if we found intelligent civilizations in space, it would confirm that this is the case.

A Consensus

Now that we have some background let's go back to our problem: How many advanced civilizations are there in space? Let's start with life in general. We have seen that it appears to be relatively easy for lower forms of life to evolve. We find amino acids and components of nucleic acids in meteorites and comets. And we appear to have found life forms on Mars. Furthermore, we have identified organic molecules in large interstellar gas clouds. So there appears to be no problem in relation to lower forms of life.

But how difficult is it for lower forms to evolve to intelligent beings? Many things have to come together in just the right way for this to happen, and as we saw the probability of this appears to be small.

As far as alien life is concerned, there are obviously three possibilities:

1. We are alone in the universe.
2. There are other civilizations out there, but they are relatively short-lived. Most become Type I, and don't get beyond it.
3. There are many Type II and III civilizations in the universe, but they consider us inferior and have little interest in our technology. They keep themselves concealed from us. This is called the zoo hypothesis.

Let's consider these, one at a time. The first is that we are alone. This is hard to accept considering the number of stars in our galaxy—200 billion—which is roughly equal to the number of grains of sand on all the beaches of the world. Is it possible that among all these stars there are no other civilizations? As we have seen this is a possibility. Despite this, though, it is still important to carry out the SETI program. Only through it can we establish if we are, indeed, the only advanced civilization, and if so, it is even more important for us to survive as long as possible.

The second hypothesis is that there are only a few Type I civilizations out there. As we have seen there are a lot of hurdles to overcome for a Type

I to advance to Type II, and perhaps even more to go from Type II to Type III. Many civilizations, even very old ones, may be so tied up with the problems of society that they haven't got time for space travel and colonization. If they are going to be long-lived there is pressure for expansion (to renew their resources), but expansion within their system, and perhaps to the planets of a few nearby stars, may be sufficient.

If there are only Type I civilizations out there, they will be sparse. A lack of Type II and III civilizations indicates that those of Type I are relatively short-lived. And if this is the case, the nearest civilization is likely to be hundreds of light-years away.

Our third hypothesis is the so-called zoo hypothesis. This assumes that there are a large number of Type II and III civilizations, but they consider us to be primitive—like ants. They may be watching us, checking on us occasionally, but have no interest in our knowledge or technology. They may be concealed from us because they are using all the energy from their star and stars around them. If so, they might be difficult to detect, but we would still be able to see them as infrared sources. They are usually referred to as Dyson spheres.

If this is the case, it is quite likely they have visited us in the past. Perhaps they check on us every hundred years or so. They would be von Däniken's "ancient astronauts." So far, though, we have seen no evidence that they have visited us in the past.

It's easy to see that there is a problem if an advanced civilization has been checking on us occasionally in the past. If we let the Earth's 4.5-billion-year history be represented by a single year, we have been civilized for only about five minutes of that year and have only had technology for less than one minute. If they visited us routinely, checking on our advances, they would have to check during one minute out of the year.

This brings us back to the question: Are we alone? We have learned a lot over the past few decades about the possibility of alien life, but we still do not know if there is life, particularly intelligent life, beyond Earth. If we don't continue searching, however, we will never know.

chapter 12

The Future: Starflight

As the Earth becomes overpopulated and its resources dwindle, humans will look to space for expansion and for new resources. It's not a question, however, of being forced into space—space colonies will never solve our population problems—it's a matter of wanting to go. Mankind has always had the urge to explore, to see what is on the other side of the mountain, the other end of the ocean. This same urge will drive humans out into the solar system—to colonize the moon and Mars, and build space stations— and finally to the stars where we may one day come face-to-face with the first aliens.

O'Neill Colonies

One of the first to advocate expansion into space was Gerard O'Neill of Princeton University. Born in Brooklyn, New York, in 1927, O'Neill obtained a Ph.D. from Cornell University in 1954. Although he specialized in high energy particle physics, he had an early interest in space and space exploration. (He was one of the scientists in NASA's short-lived scientist–astronaut program in the 1960s.) In the late 1960s, O'Neill was teaching a freshman physics class that included a section on space exploration. Curious about what the class reaction would be, he asked: Where is the best place to set up space colonies? The options were, of course, the moon, Mars, and a space satellite or habitat. He wasn't surprised when space habitats won hands down; he had, in fact, already come to that conclusion himself. But now he began thinking about how such a colony could be set up. Would it be possible? How difficult would it be? How many people could such a colony hold? What would life be like in the colony?

He decided to look into the building of a space habitat that could be used as a permanent home for a few thousand colonists. His first consideration was where it should be located. Many years earlier the French mathematician Joseph Louis Lagrange had shown that there are five points in the Earth–moon system where the gravitational pull of the earth is equal to the gravitational pull of the moon, in other words where gravity is balanced; they are referred to as libration points. One of these points is directly between the Earth and moon, and one on the other side of the moon; the others are at various positions in the moon's orbit. O'Neill decided that a libration point would be the best place for the colony.

The major problems in setting up a colony would be protection from cosmic rays and the creation of an artificial gravitational field. At the time little was known about the effects of cosmic rays, but astronauts had seen strange flashes in the dark, and it was eventually determined that they were caused by cosmic rays passing through their eyeballs. Little was also known about the prolonged effects of weightlessness. In 1973, however, *Skylab* was put in orbit (two years earlier the Russians had also put the first of their *Salyut* stations in orbit) and a considerable amount about both weightlessness and radiation protection was learned from it.

Despite the interest in *Skylab*, there was little enthusiasm for O'Neill's ideas. They were too grandiose, too far beyond *Skylab*. But this didn't deter him; he went ahead with his design and in 1974 he completed plans for a large permanent colony. He published them in 1977 in a book titled *High Frontier*. His habitat was a large cylinder that could house up to 10,000 colonists. Artificial gravity would be created by rotating the cylinder. His ideas eventually began to attract media attention. Several large newspapers and magazines printed stories about him and his proposed colonies.

Finally NASA began to take an interest, and in 1975 it called a conference to look into the possibility of space colonies. O'Neill and several other space enthusiasts attended, spending several weeks poring over the problems. The two major problems were still weightlessness and protection from cosmic rays.

With results from *Skylab*, weightlessness was now known to be a serious problem. Within days in space muscles begin to deteriorate, the

Future space colonies.

heart weakens, and bones begin to lose mass. An extensive exercise program can delay the effects, but it cannot overcome them. If people were to
live permanently in a space colony, artificial gravity would be a necessity.
O'Neill created gravity by rotating his cylinders, but there were problems
with his design. His cylinders had a relatively small radius and had to be

rotated three times a minute to simulate Earth's gravity. At such a high rate the fluids in the inner ear (that are responsible for balance) would be affected, and the colonists would become dizzy. This could be avoided by making the radius larger, since centrifugal force (the force giving the artificial field) depends on radius. A huge cylinder or sphere would not do; the group found that a torus (like a donut) was more satisfactory. With a torus, Earth's gravity could be simulated with a rotation of only about 1 rev/min. Interestingly, Werner von Braun, the German rocket scientist, had considered torus colonies several years earlier. He had even realized that shifts in weight caused by the colonists moving around would cause instabilities and had designed a system of ballasts to compensate for them.

The other problem was cosmic rays. They would have to be taken care of by the habitat itself. Soil from the moon would give considerable protection. The outer part of the torus—where the artificial gravitational field is the strongest—would be covered with lunar soil to a depth of six feet. Cosmic rays would not be able to penetrate it. But in the "upward" direction—in other words, the inner part of the torus—there would have to be large windows to let in sunlight. The windows would give some protection, but sunlight would have to be directed into the torus so a large mirror would be needed over the top of the torus. It would give additional protection.

The habitat would be constructed in space with most of the materials being brought from the moon. The Earth has a much stronger gravitational field than the moon, and considerable energy would need to be expended to bring material from it. Furthermore, the moon has no atmosphere and there would be no friction in launching from it.

The Apollo program gave the NASA group a good idea of what was available from the moon. The lunar soil is rich in iron, aluminum, titanium, and magnesium; of particular importance it also has considerable oxygen (40 percent) and silicon (20 percent). The oxygen could be used in the atmosphere of the habitat, and also it could be combined with hydrogen for water. Unfortunately, there is little hydrogen on the moon, so it would have to be brought from Earth. The silicon would be necessary for the construction of the large windows that would be needed.

What would it be like to live in such a colony? According to the design that was finally agreed upon, residential areas would be alternated with agricultural and wooded areas, or parks. The residential areas would consist mainly of apartment houses and crops would be raised in the agricultural areas; animals could also be raised in the agricultural areas.

Purification of the air and water would be vital, as would recycling. But aside from a few things that would have to be brought from Earth, the colony could be made self-sufficient. Furthermore, it could be of considerable help to Earth. Conversion of solar energy is much more efficient in space than it is on Earth, so solar energy stations could be set up for generating power for Earth.

To the Stars

As the O'Neill colonies became larger and more sophisticated they would move farther and farther from Earth. Eventually they would leave Earth orbit and take up residence around the sun. A region that would no doubt be of particular interest to them is the asteroid belt. Many of the colonies would take up residence here because of the tremendous resources. The next step would be the vast reaches beyond the solar system—trips to nearby stars. But as we will see, that is a tremendous jump, and will not be easy.

The first destinations would be nearby stars such as Alpha Centauri at a distance of 4.3 light-years, Barnard's star at 5.9 light-years, Epsilon Eridani at 10.8 light-years, and Tau Ceti at 12.2 light-years. It is easy to see, however, that there are problems. If we used an *Apollo* spacecraft to go to Alpha Centauri it would take 850,000 years to get there, and the same amount of time to get back. Even the faster *Voyager* spacecraft that is now headed for the stars would take 80,000 years to get to Alpha Centauri. It's obvious that we will need spaceships that are capable of much greater velocities if we are to reach the stars. But even if we had them there would still be a problem: according to relativity we are restricted to the speed of light. This means that if a star is 20 light-years away there is no way we can get to it in under 20 years. The overall trip—there and back—would take at least 40 years, and probably much longer.

Let's assume that we can travel close to the speed of light. What would a trip to Alpha Centauri be like? After taking off from Earth we would have to accelerate to our top speed; this would take at least a year because of the gravitational, or *g* forces. (High accelerations cause high *g* forces.) We could then coast for a while, but as we approached Alpha Centauri we would have to decelerate for a year to get to a speed low enough to observe (or possibly land on) a planet orbiting the star, if it had one. The trip back would be similar: we would accelerate to top speed, coast, then decelerate as we approached Earth. The overall trip would take at least 12 years, and Alpha Centauri isn't a particularly good candidate for life. Most of the good candidates are much farther away.

In addition, the time needed for the trip isn't the only difficulty; fuel is even more of a problem. A convenient measure of the fuel needed is called the "mass ratio"; it is the mass of the starting rocket (which is mostly fuel) to the payload (mass at the final destination). The mass ratio for the trip to Alpha Centauri at a speed close to that of light is approximately a billion. In other words, for every ton we wanted to get back to Earth, we would need a billion tons of fuel. This is obviously far beyond anything we are capable of.

Another problem is the force on the crew's bodies during acceleration and deceleration. The human body is designed for approximately 1 g (the gravity of Earth) and it can't take much over this for any length of time. Astronauts and pilots can withstand 15 g's for a few seconds, but beyond that they blank out. Even a few g's over a period of days would be extremely uncomfortable. Furthermore, the weightlessness of the coasting stage would also be a problem. As we saw earlier, weightlessness causes all kinds of difficulties so there would have to be some gravity in the spaceship throughout the trip.

With our advancing technology, rocket speeds are increasing rapidly, but it's unlikely we'll reach speeds anywhere near the speed of light during the next century. A more realistic goal is about 10 percent the speed of light. Let's consider spaceship designs that would be capable of this.

Fission and Fusion

Chemical fuels are obviously out; they give speeds that are only a small fraction of the speed of light. We therefore have to turn to nuclear-powered engines, and here we have two choices: fission and fusion. We have had fission reactors for years and have considerable experience with them, but they would be incapable of pushing a spaceship anywhere near the speed of light. Plutonium and uranium in liquid form have much hotter burning cycles and would produce considerably more energy. Speeds of a few percent that of light would be possible with such an engine.

A different approach was taken in 1958 in what was called Project Orion. Involved in the project were Stanislaus Ulum, Ted Taylor, Freeman

Dyson, and others of Los Alamos Scientific Laboratory. The propulsion system they worked out was centered around hydrogen bombs. There was, at the time, a huge stockpile of hydrogen bombs, and they thought it would be a good use for them. One of the major components of their design was a large, curved shield that could withstand an atomic blast; it was attached to the rear of the spacecraft. In the center of the shield was a hole through which hydrogen bombs could be fed. They would be detonated a short distance behind the shield, and the momentum from the blast would push the spaceship forward. According to their design, a bomb would be detonated every 100 seconds. The spaceship would obviously have to be cushioned from the concussion, and a shock absorber was designed to do this; it was placed between the spaceship and the shield. Dyson and his colleagues worked on the project for several years, but with the test ban treaty and other problems it was finally cancelled in 1965.

An alternative to that design is a fusion reactor within the spaceship. So far, however, we have been unable to build a working fusion reactor— at least not one that we get more energy out of than we put in. In the fusion process there is a "joining," or fusing, of nuclei; in the sun, for example, four hydrogen nuclei fuse together to form a helium nucleus. The fusion process is achieved by applying enough energy so that the nuclei get close together. The main advantage of fusion, aside from the fact that it supplies more energy per unit mass than fission, is that in its simplest form it requires hydrogen, which is abundant in the universe, whereas fission requires heavy elements, which are rare. The main problem with it here on Earth is that it is difficult to contain. The temperature at which fusion takes place is so high there is no known material that can contain it. Magnetic fields have been used, but there are difficulties with them. It is our hope,

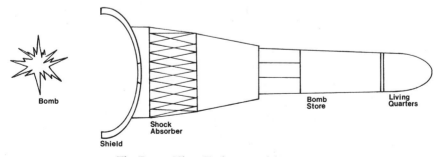

The Dyson-Ulum-Taylor propulsion system.

however, that these problems will be overcome in the next few years, and fusion reactors will be built.

The most common form of hydrogen on Earth is a gas, but like all substances it can also exist as a liquid or solid. The most promising form for the engine of a spaceship is metallic hydrogen. Scientists at the Carnegie Institution of Washington believe they may have produced this rare form when they subjected hydrogen to a pressure of approximately 1 million atmospheres. But it will no doubt be some time before it is manufactured routinely.

Research also indicates that fusion reactions can be achieved using small pellets of deuterium and helium-3 (a light form of helium). In this case a pellet is bombarded on all sides by powerful laser beams. The beams heat the pellet (to 100 million degrees) and compress it until fusion occurs. The lasers needed for such a device are still far beyond what we have, but they may be available in the next few decades.

For a spaceship to achieve the greatest possible velocity, the velocity and mass of the exhaust gases have to be as high as possible. In the case of a fusion reaction, a plasma is created that can be guided through an exhaust nozzle using magnetic fields. The velocity of this plasma would be high enough to give the spaceship, over time, a speed approaching 10 percent that of light.

The fusion engine was designed into the British spaceship *Daedalus*. The Daedalus project was initiated in the early 1970s by scientists of the British Interplanetary Society. They wanted a spaceship that could get to Barnard's star within a human lifetime and decided that a pulsed fusion reaction using lasers was the most appropriate energy source, with pellets being detonated at the rate of 250 per second. Accelerating at this rate for four years, the spaceship would acquire a velocity 12 percent that of light, and would take about 50 years to get to Barnard's star.

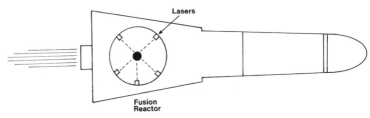

Rocket with fusion reactor.

Current technology was assumed in the Daedalus project, but there were problems. Considerable helium-3 would be needed for the pellets and it is rare on Earth; most of it would have to be obtained from the moon. Furthermore, the *Daedalus* spaceship was assumed to be unmanned; a manned version would require much more fuel, and the journey would take considerably longer because the rate of acceleration would have to be considerably lower.

Bussard's Ramscoop

One of the major problems with the spaceships discussed so far is that most of the fuel has to be carried on the trip; furthermore, there's a catch-22. The bigger and faster the spaceship, the more fuel that is needed, but it's the fuel that makes up most of the weight. In 1960 R. W. Bussard of Los Alamos Scientific Laboratory proposed an ingenious way around this problem. He suggested that a huge "ramscoop" could be used to collect hydrogen and helium from interstellar space. Dust and gas fill interstellar space with an average density of approximately 3 atoms per cubic inch, and some of it would be appropriate as fuel. The ramscoop would collect this material and use it in a fusion reactor. The amount of gas that would have to be collected is enormous and the scoop would therefore have to be huge. It would consist of a gigantic magnetic field about 2000 miles across.

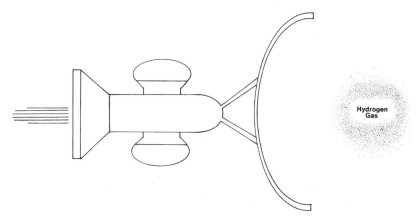

Bussard's ramscoop. In practice, the scoop at the front of the rocket would have to be hundreds of miles across.

A spaceship using a ramscoop could reach 10 percent the speed of light without too much difficulty, and at this speed a trip to Alpha Centauri would take only about 40 years. There are, however, several difficulties with the concept. Another energy source would be needed for the initial acceleration, in other words, to get the spaceship up to speed, as the ramscoop would not be efficient at speeds much under 10 percent that of light. Also, the particles being collected would give off considerable radiation, and the huge magnetic fields needed would be a problem.

Interstellar Sails

Another way around the problem of carrying fuel is to use a "solar sail." In this case a sail—perhaps as large as 150 miles across—would be needed. It would be pushed by a beam of light from Earth. Light exerts pressure and would act like a "wind." An ordinary light beam disperses too rapidly to be of any use, but a laser beam can be projected for millions of miles into space without significant spreading. With a sufficiently strong laser, the ship could be accelerated to speeds as high as 30 percent that of light over a period of about ten years.

The major problem with the idea is the intensity of the beam that would be needed. It is far beyond the capability of any lasers we have on Earth at the present time. In practice, it would take thousands of lasers in orbit around either the Earth or the sun to produce the needed intensity. The beams from these lasers would have to be focused on the sail using a large lens that was in orbit.

In addition, the method would be limited to the nearest stars, and there would be no way of slowing the spaceship down when it got to its destination.

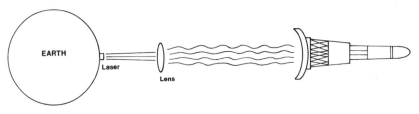

Interstellar sail using laser.

An alternative to the laser sail is the microwave sail. It could be made of lightweight mesh; the agent providing the pressure in this case would be microwaves. According to present estimates it would be as effective as the laser sail.

The Matter–Antimatter Spaceship

In theory the most efficient rocket ship would be one powered by matter–antimatter annihilation. In the early 1930s Paul Adrien Maurice Dirac of Cambridge University showed that corresponding to each type of particle there was an antiparticle, usually of opposite charge. Corresponding to the electron, for example, there was a positively charged antielectron (now called the positron), and to the proton, a negatively charged antiproton. And if you brought a particle and its antiparticle together, they annihilated one another with the release of considerable energy in the form of photons. This is conversion of matter into pure energy, and in theory there is no way of getting more energy.

The German rocket scientist Eugen Sanger proposed a matter–antimatter engine in the 1950s but he didn't come up with a working model, so there was little interest in his suggestion. In the early 1980s, however, Robert Forward of Hughes Research Labs became interested in the idea. Taking a year's leave of absence from Hughes, he was able to produce a working model.

There are two problems with antimatter: containing it and producing it. Antimatter is difficult to contain because it is annihilated when it comes in contact with matter, so it must be kept isolated. Scientists do this using magnetic fields. Charged particles trace out circular paths on magnetic fields, and a beam of antiparticles can therefore be kept in circulation at the center of a torus using electromagnets. Unfortunately, at the present time they can be stored in this way for only a few days.

The second problem is obtaining the antimatter. From our study of cosmic rays we know there is little antimatter in space, so it would have to be produced either on the spaceship, or on Earth prior to flight. Antimatter is not difficult to produce; it is done every day in the large accelerator at Fermilab, but it is expensive. It can be produced by projecting a beam of protons at a lithium target. One of the products of the reaction is antiprotons that are easily collected and stored using magnetic fields. The problem is that the energy needed to produce them is greater than the energy released when they are annihilated. In other words, the energy required to produce them is a million times the annihilation energy; however, it is expected that this will improve over the next few decades. Velocities close to that of light could be reached with matter–antimatter spaceships, so despite the difficulties they hold a lot of promise.

Another possibility related to this is a spaceship based on negative matter. Negative matter is different from antimatter in that it produces antigravity, which means that matter would be repelled from it. There's nothing that forbids the existence of negative matter, but so far we have no indication that it exists. Since matter and negative matter repel one another a spaceship could easily be designed using the resulting force, but we're still not sure how it would be controlled.

Cryobiology and Hibernation

Near the beginning of the science fiction book *Alien*, the crew members wake from a state of "hypersleep." They are in chambers, covered with a "preservative cryosleep fluid," a fluid that has presumably kept them in a suspended state for many years, a state in which they did not age.

Is such a thing possible? As you might expect, the idea has been used extensively in science fiction, and it is an alternative to high speeds. If you want to cross the large distances between stars, you either have to travel at speeds close to that of light (which we have seen is difficult), or you have to have long spans of time pass without it affecting your body. The logical way to do the latter is to go into a "deep sleep"—a form of hibernation that might last for decades or even centuries. This would, of course, require an elaborate computer system to manage the spaceship, monitor and nourish the people in the deep sleep, and finally to wake them at the right time.

Hibernation is, of course, common in the animal world. Bears, squirrels, and others go into a deep sleep where their body temperatures fall to just above freezing, and their heartbeats decrease to only a few per second. During this time their bodies age very little. Scientists have been studying this state in an effort to see what causes it. Is there an enzyme in the animal that comes into play? Does a particular gene cause the phenomenon? If scientists eventually find out, it may become possible to genetically engineer humans to go into hibernation.

Even better, perhaps, would be freezing. It is known, for example, that fish can be frozen and revived. But a considerable amount of study has been done in this area (cryobiology) and the prospects don't look good. The body is made up mostly of water in cells, and it's well-known that when water freezes it expands. This causes cell walls to rupture. Furthermore, when frozen cells thaw they buckle and bend in the same way paved streets do during the spring thaw. Many cells can replace themselves, but some—in particular, brain cells—cannot.

An alternative to freezing humans is freezing embryos. It may be possible to keep them in a frozen state during the flight, then thaw them and bring them to maturity at the destination.

Another possibility is that as the body ages, parts of it could be replaced by computerized synthetic parts—something like the $6-million man of the television show. There are, indeed, many possibilities, but most of them are still far in the future.

Relativistic Time Travel

A well-known way of slowing down time comes from the theory of relativity. Einstein showed that time is not the same for everybody; it depends on their relative speeds. When two astronauts, for example, are in relative motion, the time that passes on their watches is not the same. The easiest way to explain the effect is to consider two clocks, one attached to the Earth, and one on a spaceship that leaves Earth and travels to a star at high speed. Let's consider speeds of $0.5c$, $0.99c$, $0.999c$, where c is an abbreviation for the speed of light. Assume the trip to the star takes one year according to the clock aboard the spaceship. We then ask: How much time passes back on Earth? If the speed of the spaceship is $0.5c$, 1.1 years will pass back on Earth—not a significant difference. If the speed is $0.99c$, however, 7 years will pass back on Earth, and at $0.999c$, 22 years will pass. This means that for twins, each 25 years old at the beginning of the trip at $0.999c$, one would be 26 at the end of it and the other 47.

It's easy to see from this that if we travel fast enough, particularly if we travel at speeds close to that of light, we can shorten the trip considerably. If we went back to Earth, however, many years would have passed. Still, in theory, we could get to nearby stars in only a few years, or perhaps much less.

There is, unfortunately, a catch. Not only does time slow down at high speeds, but masses increase. In short, the spaceship gets heavier and heavier as it goes faster and faster relative to Earth, and since it weighs more, it takes more energy to accelerate it. The amount of energy required to accelerate it close to the speed of light is huge, to say the least; it is far beyond anything we're likely to be ever capable of. In practice, a speed of about 87 percent that of light is optimal for space travel.

Tachyons and Warp Speed

A concept you read about frequently in science fiction is warp speed. Warp speeds are speeds beyond the speed of light. But as we saw, relativity theory tell us that matter cannot travel at the speed of light. Looking closer, however, we see that it does not forbid speeds beyond that of light.

Gerald Feinberg of Columbia University looked into what happens beyond the speed of light. He found that Einstein's equations became imaginary, but that was not a problem; mathematicians had been dealing with imaginary quantities for years. (There's a whole branch of mathematics called complex analysis that deals with it.) To us, the "world" beyond c would be imaginary, but in theory, particles could exist there. Feinberg called them tachyons; these tachyons could travel only at speeds greater than that of light. Scientists have looked for Feinberg's tachyons without success, but it is possible they haven't looked hard enough. As unlikely as it seems, it is possible that this "world of tachyons" exists, and if we could get to it, we would be able to travel at speeds greater than that of light. This is the warp drive of science fiction.

There is still the problem, however, of overcoming the speed of light. We can't travel at c, and if we want to get to warp speeds we have to pass it. One possibility for doing this is a "quantum jump." It's well-known in quantum physics that a particle can move from one position to another without actually traversing the space between the two positions; the particle just appears at the second position after a quantum jump. In the case of a spaceship, we would have to take it up to $0.999c$, then quantum jump to $1.01c$, where we would then be able to travel at speeds several times that of light.

Wormholes

Perhaps the easiest way to cover large distances across our galaxy, or the universe, would be to use a "shortcut" through space. For years it was

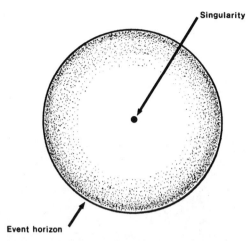

Singularity

Event horizon

A black hole showing event horizon and singularity.

believed that black holes might give us this shortcut. Black holes are the remnants of collapsed stars; they are usually a few miles across and consist of a surface called an event horizon with a singularity at the center. Once anything passes through this event horizon it cannot get out; it is, in essence, a one-way surface. Leading up to the event horizon is a curved region of space referred to as a wormhole. Science fiction writers have used these wormholes as bridges to distant parts of the universe for years, but scientists knew there were problems associated with them and it was unlikely they would ever be used in this way. The main problems were stretching forces that pulled you apart when you got close to the black hole, pulsations that cut off access, and radiation from the singularity. Furthermore, they were one-way; you could go through in one direction, but not the other.

In the early 1980s the late Carl Sagan of Cornell University wrote the science fiction novel *Contact* (which has since been made into a movie). In the novel he had his heroine Ellie Arroway detect a message from space. When decoded it was found to be the instructions for building a "machine" that could take people to a distant planet. In the original version of the book Sagan used the wormhole associated with a black hole to transport Ellie and several colleagues to the planet, but as a scientist he wanted to be sure the science in the book was as accurate as possible, so he sent the manuscript to Kip Thorne of Caltech, one of the foremost experts on black holes in the world.

When Thorne read the manuscript he was disturbed to find Sagan had used the wormhole associated with a black hole; he knew there were many problems associated with them. Nevertheless, he wanted to help Sagan so

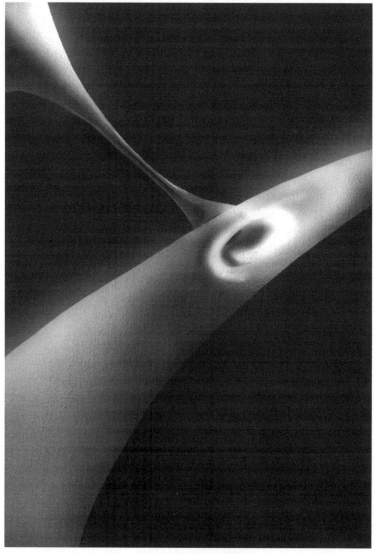

A wormhole in space.

he sat down and took a look at Einstein's equations. Was it possible that a solution had been missed that would allow a shortcut through space? To his surprise he found one. If a wormhole existed independent of a black hole, it could be made into an interstellar subway. All the problems mentioned earlier could be overcome. He would have to cover the inside of the wormhole with "exotic material" to stop it from pulsing, but this was something that might eventually be possible. Furthermore, the wormhole could be made relatively small—only a few feet across if necessary—while the wormholes associated with black holes were miles across.

Thorne left many of the details to his student Michael Morris to work out. Morris and others eventually discovered that these wormholes could be made into time machines. Not only could you enter one and end up at a distant part of the universe seconds later, you could also go backward and forward in time. In theory, if both the entrance and exit of the wormhole were on Earth, you could enter the wormhole and come back to the Earth hundreds of years in the past or future. This is, of course, common in science fiction, but it was the first time it was shown it might be possible in fact.

Wormholes such as this would be the ideal way to traverse the tremendous distances between the stars. We wouldn't have to worry about ultrafast spaceships or deep sleep. If this is, indeed, a viable alternative, supercivilizations such as Type IIIs may even now have a network of subways throughout our galaxy. Is there any way we could detect such a network? Unless there was an opening close to Earth, it is quite unlikely. It's not something you could see with a telescope; it's merely curved space.

If we did find firm evidence of a Type III civilization in a nearby galaxy, it would, perhaps, tell us that a network of wormholes is possible. Communication and travel via ordinary routes over a large fraction of a galaxy would be so difficult that it's virtually impossible a Type III civilization could be under central control without such a system. Messages would, in some cases, take hundreds of thousands of years to get to their destination.

Other Dimensions and Parallel Universes

Higher dimensions, parallel universes, and other universes may seem like highly speculative concepts, but they are seriously considered in modern physics. Researchers working on the early universe and the "theory of everything" commonly use higher dimensions. One of their theories has as many as 26 dimensions. But we can observe only 3 dimensions of space and 1 of time. What is the significance of the other dimensions? According to scientists, they are "curled up" so that we can't observe them. Could we make contact with such a higher dimensional world? It's highly unlikely, but it can't be ruled out. It might be possible, for example, to pass somehow to this higher dimensional world, use it to traverse long distances across the universe, and then pass back into our universe. We have no idea how this might be done, but it's an interesting thought.

We can ask the same question about "parallel universes." This concept comes from quantum theory. According to the theory, as time passes, our universe breaks off into several parallel universes. Again, we have no idea how we could contact these parallel universes, or even if they exist for sure, but they are fascinating.

Finally, we have the "other universes" of cosmology. They come from inflation theory. According to this theory, our universe is only one of many; the others, however, are a long distance away, and even if they did exist it seems unlikely we could ever be able to make contact with them. According to present theories they are forever cut off from us.

Most of the ideas presented in this section may seem beyond reality, but there is no doubt that there is much about the universe we still do not understand.

chapter 13

Have We Been Visited?
UFOs and Ancient Astronauts

In February 1974, two brothers left Idaho with a truck full of furniture and headed for Ely, Nevada. They were about 100 miles down the road when the driver spotted a bright, blue-green light off to the left of the road. He woke his brother, who was sleeping at the time, and they both watched the object for several minutes. Then other similar objects appeared and began to follow the truck. One of the objects separated from the others; it moved overhead in front of the truck, then disappeared.

Suddenly the truck lost power, then the steering went and the driver lost control, but after a long struggle he managed to get it stopped. Jumping out to see what was wrong, he was startled by a glowing object covering the road in front of him. It appeared to be moving toward him. In panic he jumped back in the truck and locked the doors. As the two men stared at the object it suddenly disappeared.

In October 1973, a truck driver and his wife were traveling near Cape Girardeau, Missouri, in the early hours of the morning. The man noticed a strange, shining red-and-yellow object in his rearview mirror. As it got closer he could see that it was turnip-shaped with three sections, one above the other. It seemed to be made of aluminum or chromium. Sticking his head out of the window for a better look, he watched the object move directly over his truck. Suddenly a bright beam shot out of the craft. He screamed and clutched his eyes. When he finally got the truck stopped, his wife switched on the cab lights and saw that his face was red and covered with sweat. One lens of his glasses was missing and the frames were strangely twisted—almost as if they had been melted. She rushed him to the hospital. Five days later he still had only 20 percent vision.

In November 1973, a woman in New Hampshire was driving home about 4:00 A.M. when she noticed a strange blinking light overhead. At first she paid little attention to it, but she soon realized it was getting closer. It was following her! She began to panic. As it approached her car, she could see the object clearly. It was round with a strange hexagonal design on the side, and through a small oval window she saw a figure moving about. It had large, strange eyes and gray, wrinkled, leathery skin. A high-pitched, whining noise was emanating from the craft. Fearing for her life she swerved into the nearest driveway, jumped out of the car, ran up to the house, and pounded on the door. When the owner got to the door, he found a sobbing, almost hysterical woman. At first the man saw nothing, but later both he and a policeman saw a colored light in the distance that blinked on and off just as she had described.

In April 1964, near Socorro, New Mexico, a deputy sheriff was pursuing a speeder when he heard a loud noise and saw a bright flash to his right. Thinking it was an explosion or fire, he abandoned the chase and headed toward it. Over the hill he saw a strange craft lowering itself to the ground; a bright flame shot out from beneath it. He radioed headquarters as he approached it cautiously. It was about 15 feet long, oval, with no doors or windows, and it appeared to be made of shiny material, possibly aluminum. He was about 20 feet away from it when suddenly bright flames appeared again and he jumped back. As he watched, it rose from the ground and disappeared. Walking over to where it landed he saw marks on the ground where it had depressed the soil and burned the vegetation beneath it.

In October 1973, two shipyard workers were fishing near the mouth of the Pascagoula River in Mississippi when they heard a low buzzing sound from behind them. Turning, they were astounded to see a glowing object hovering just above the ground only a few feet away. Suddenly three grayish humanoid creatures with bullet-shaped heads appeared. They paralyzed the two men and floated them aboard the craft. Inside, one of the men reported that he was examined by a strange electronic eye. They were in such a weakened condition when they were released that they could hardly stand. At first they decided not to relate their story, sure that no one would believe them. But later, talking it over, they went to a reporter and told him. The event was national news within days.

These stories may sound like excerpts from old science fiction movies. Were they hoaxes? Consider the two fishermen. Astronomy professor and UFO investigator Allen Hynek was convinced after interviewing them that they had, indeed, undergone a traumatic experience. He said that they were not good enough actors to pull off a hoax. In addition, they were left alone in a room with a tape recorder going (unknown to them) shortly after the incident, and there was no indication that they were trying to fool anyone. Finally, they passed a lie detector test.

What about the others? The woman from New Hampshire would hardly talk about her experience only a few weeks later. The Missouri truck driver lost part of his eyesight for several weeks. And the truck drivers on their way to Ely, Nevada, kept their identity secret. None of them became rich as a result of their experience.

These are only a few of the thousands of UFO cases reported by average, law-abiding people who appeared to honestly believe what they saw and had little to gain in the perpetration of a hoax. And in most cases they had a lot to lose. People who claim to have seen UFOs are frequently shunned and laughed at. That's not to say, however, that there haven't been hoaxes. There have, indeed, been many (they will be discussed in the next section).

The First Flying Saucers

The modern era of UFOs is considered to have begun with the sighting of several "flying saucers" by Kenneth Arnold on June 24, 1947. Arnold, an air rescue pilot from Boise, Idaho, was flying to Yakima, Washington, over the Cascade Mountains. A plane had been reported missing in the area and he was looking for it. As he scanned the mountain peaks, a bright flash attracted his attention. Looking toward it he saw nine bright objects flying near Mt. Rainier. They appeared to be in formation and were flying with a tremendous speed; he estimated it to be 1700 mph (the top speed of an aircraft at that time was approximately 700 mph).

When he landed at Yakima he told the ground crew what he had seen, admitting it was the most bizarre thing he had ever observed. The news leaked out and he was soon besieged by reporters. He told one of them that the objects were like "shining saucers" that acted like flat rocks skipping across water. The reporter grabbed onto the words "flying saucer" and it has been part of the folklore ever since. To this day Arnold is still not sure what he saw, but he later mentioned that the objects didn't look like saucers to him; they appeared to have wings. Within a short time there were numerous other sightings (the total for 1947 was 850).

Arnold got so much publicity over the incident that he later said he was sorry he had reported it and swore he wouldn't report anything like it again, if it ever happened to him.

The 1947 "wave" of reports was the first of many. For some unknown reason, reports seem to come in waves; there were waves in 1947, 1952, 1965, and 1973–1974, and as historians have looked back in time they have found that many waves and sightings occurred well before Arnold's sighting in 1947. They were even mentioned in the Bible. According to Ezekial: "And I looked, and behold, a whirlwind came out to the north, a great cloud, and fire infolding itself, and a brightness was about it … out of the midst thereof, came the likeness of four living creatures. And this was their appearance, they had the likeness of a man."

Columbus is said to have spotted a UFO from the deck of his ship *Santa Maria*. He described it as "a light glowing at a great distance." He called several of his crew and they witnessed it with him. UFOs were reported in ancient India and in Tibet. Indeed, the 1947 wave in the United States was not the first one here. A large number of UFOs were sighted in California near the turn of the century.

The UFO phenomenon is still with us. Files now exist that contain close to 100,000 recorded cases. With such a large number, it's obviously something that's not going to go away. And although most scientists are reluctant to be associated with it, many now believe it is worthy of detailed study. One of the problems of UFOlogy is the large number of hoaxes that have been perpetrated. The most famous of the early hoaxers was George Adamski. He claimed to have visited several nearby planets in UFOs, to have met beautiful space women and to have discussed cosmic matters with space rulers. From his base of operations—a small cafe in California— he sold photos of UFOs and various other space paraphernalia. Without realizing how much astronomers knew about the moon, he even claimed to have visited lush cities on it. His two books, *Flying Saucers Have Landed* (1953) and *Inside the Space Ships* (1956), enjoyed relatively good sales. Other well-known hoaxers were Daniel Fry, Truman Bethurum, and Howard Menger, all of whom claimed to have had friendly discussions with UFO occupants and to have flown in UFOs. The modern tabloid press also

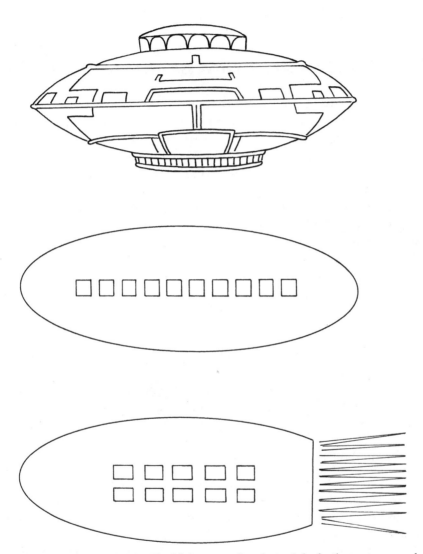

The three most common UFOs. The "flying saucer" at the top is by far the most commonly sighted.

gives the phenomenon a bad name with headlines such as "UFO Lands on Carrier," "UFO Baby Found on Mt. Everest," "UFOs and the Bermuda Triangle," "UFOs and Bigfoot," and so on.

The Classification of UFO Reports

One of the first scientists to associate himself with the UFO phenomenon was Allen Hynek. A professor of astronomy at Ohio State University and later chairman of the astronomy department at Northwestern University, he was hired by the Air Force in 1949 as a consultant. Initially, he was a skeptic, and when I talked to him in the early 1970s he was still skeptical of most reports, but he believed there were a small number that were worth detailed study. He emphasized that most UFO reports did not come from fanatics or UFO buffs; most came from average citizens who had little interest in the UFO phenomenon prior to their sighting. Some, he stressed, even come from scientifically trained people (in most cases they remain anonymous). In 1972 he published *The UFO Experience: A Scientific Enquiry* which has become a standard reference in the area.

Hynek classified close encounters into three types (we'll ignore distant sightings for now):

1. First Kind: The UFO is seen at close range. Its size and shape are distinguished.
2. Second Kind: These are similar to the first kind but there is some interaction with the environment. Vegetation is left flattened, the ground is scorched, or the engine of a nearby car is stalled.
3. Third Kind: This is the case where an "occupant" is seen near or in the UFO.

More recently another classification has been added to these; it is referred to as a close encounter of the fourth kind. In this case the witnesses are abducted, taken aboard the UFO, and examined. The case of the two fishermen at Pascagoula is of this type.

Close encounters of the fourth kind may be hard to accept or believe, but there have been so many of them in recent years that a conference on the phenomenon was called at MIT in June 1992. And it was not called by UFOlogists, or space fanatics, but by two professors with impeccable credentials. They were David Pritchard of MIT and John Mack of Harvard. Pritchard is a physicist who has done outstanding work in atomic, molecular, and optical physics; he was awarded the prestigious Broida Award in 1991 for this work. Mack is a graduate of Harvard Medical School and formerly head of the department of psychiatry at Cambridge Hospital, Harvard Medical School. He was recently awarded a Pulitzer Prize for a book on Lawrence of Arabia.

Arguments for Extraterrestrial UFOs

There have always been, and always will be, a large number of people who can only be classified as true believers; they are die-hards who are convinced that UFOs are from another world and there is very little you can do to change their minds. They give many arguments for their point of view. One of their strongest arguments is the large number of reports that have occurred over the years; it is now probably in excess of 100,000, depending on what you count. Another argument they give was presented in an earlier chapter. In chapter 11, "Are We Alone?" we saw that if only one civilization advanced to the state of a supercivilization, it could colonize our entire galaxy in about 100 million years, which is only a small fraction of its 16-billion-year life. And of course with the large number of stars out there—the probability seems large that at least one civilization will break through to this stage.

Another argument that UFOlogists give is, "We don't know everything. In fact we probably can't even visualize what a civilization that is a million years old would look like." And indeed this is true; about all we can really do is speculate about what it might be like, and how it might use wormholes, higher dimensions, hyperspace, or warp speeds for travel— but we can never really know for sure.

Many UFOlogists point to a government cover-up, and this is always a possibility. But to most scientists it seems unlikely. It's interesting that Jimmy Carter, during his 1976 campaign, admitted that he saw a UFO in Georgia in 1969. He said it was bright, changed color, and was about the size of the moon as seen in the sky; he claimed that he watched it for about ten minutes. He promised to look into UFOs if he was elected president. From all accounts he did attempt to open secret UFO files but was dissuaded from pushing his inquiry too far.

Arguments Against Extraterrestrial UFOs

Most astronomers do not take UFOs seriously, although a growing number believe they should be investigated more thoroughly. The major problems with the UFO reports according to them are as follows:

1. There are few, if any, good close-up photos of UFOs. (Yet cameras and video cameras are common; almost every tourist has one around her neck.)
2. There are no pictures of aliens, only drawings as people remember them.
3. There is nothing from UFOs—no mementos, nothing left as proof of their existence.
4. No UFO has ever crashed, with the remains being found (but see the following section on the Roswell incident).
5. No bodies of aliens have ever been put on display.

In short, there is no really hard evidence. Considering the number of aircraft that have crashed on Earth, it is surprising that a UFO has never crashed (with the remains being found). As you might expect, though, there have been "reports" of crashed vehicles. In 1957, for example, a "flying saucer" was reported to have exploded in the skies over Brazil. Most of the debris fell into the ocean, but a few pieces were apparently recovered in the shallows by a fisherman. According to Brazilian officials, tests showed the material recovered was magnesium of extreme purity and incredible strength, but no tests were ever allowed by persons outside of Brazil. And nothing was ever finally heard about the case. And there has been at least one reported crash in the United States, but we will discuss it later (again, the Roswell incident).

Another difficulty with the extraterrestrial argument has already been discussed in earlier chapters: the tremendous distances between stars, and the problems related to fuel and the time to get to them. UFOlogists argue that the aliens controlling the UFOs are using a mode of travel we are unfamiliar with, and of course, that is possible.

Another problem centers around the frequency of visits. If, say, only 1 percent of reported UFOs are extraterrestrial, this is still an incredible number considering what an insignificant speck we are in the universe. Earth is merely one of nine planets orbiting an average, run-of-the-mill star, which is one of 200 billion stars in our galaxy. If they are making long flights through space, we would be lucky to be visited once in 1000 years. Even if there were more than one civilization visiting us, it is unlikely we would be visited more than once a century.

Why are we visited so often as the reported sightings would indicate? Again, it can be argued that their mode of travel is far beyond anything we can imagine. Most skeptics, however, find such arguments to be a lot of nonsense. Philip Klass, senior editor of *Aviation Week and Space Technology*, is one of the best known skeptics and debunkers. He has shown that large numbers of "unexplainable" reports can be explained. He published a long list of them in his book *UFOs Explained*, which was published in 1976. Klass has gone so far as putting up a $10,000 wager that no clear proof that UFOs are of extraterrestrial origin will be found. According to his terms, anyone taking up his wager must pay him $100 each year that there is no proof. Interestingly, several people have taken him up, but so far no one has collected.

Project Bluebook and the Condon Report

One of the first in a series of scientific investigations into the UFO phenomenon came in 1952; it was called the Robertson Committee, and was headed by world-famous relativist H. P. Robertson. The committee spent 12 hours reviewing the evidence and decided there was nothing to it. The projects that followed were generally under the jurisdiction of the Air Force; they were Project Sign, Project Grudge, and Project Bluebook.

Allen Hynek was hired as a scientific consultant to Project Bluebook. Over several years he and several assistants gathered an impressive file of cases, but according to Hynek little money was allotted for detailed study of them. The Air Force eventually became dissatisfied with the progress that was being made; furthermore, it was getting considerable flak from both UFOlogists and UFO skeptics to get to the heart of the problem. So in 1966 it decided to let an independent scientific team settle the issue once and for all. The man hired to head the group was Nobel laureate E. U. Condon of the University of Colorado. Condon and his committee selected a group of cases for careful study. Their report was finally issued in 1968; it was over 1000 pages long but did little to settle the feud. Both sides— UFOlogists and skeptics—criticized it severely. The report recommended that the Air Force give up its investigation, stating: "The study of UFOs is adding nothing to the advancement of science." And indeed, as a result of the report, Project Bluebook was closed down.

Everyone expected the reports to die away after the Condon report, and for a while there was a lull. Then came the 1973–1974 wave with some of the most impressive reports up to that time. Hynek estimated that there were as many as 100 sightings a day worldwide during this wave.

The Star Map

One of the most famous UFO stories, perhaps because it involved one of the first abductions, is the case of Barney and Betty Hill. The Hills were an interracial couple, respected in their community, and prior to their incident they apparently had little interest in UFOs (Betty did have a sister who had reported seeing a UFO earlier, however). In 1961 the two were returning home to Portsmouth, New Hampshire, from Canada when they observed a bright object in the sky. After it began following them, Barney watched it with binoculars for several minutes; he thought he might have seen somebody at the window of the craft.

They began to worry when it came overhead, but after two beeping sounds, it disappeared, and they continued on home. When they got home they discovered to their surprise that there were two hours they could not account for. Over the next few months they both had bizarre dreams so they went to a psychologist who referred them to a psychiatrist, Dr. Benjamin Simon.

Simon put both Barney and Betty under hypnosis to see if he could find out what had happened. Both remembered having been taken into the UFO by small, hairless beings with large heads, but a much more detailed account of the ordeal came from Betty. She said that both were examined medically, and a large needle was inserted in her stomach. The beings said nothing, but they did communicate telepathically.

Betty said she saw a star map on the wall of the UFO while she was inside that included their home planet. Simon took her back to the incident using hypnosis, and she made a rough copy of what she had seen. (Several stars were joined by lines on the map.) Marjorie Fish, a research assistant at Oak Ridge National Laboratory in Tennessee, heard about the map and got in touch with Betty. In 1969 they got together in an attempt to see if Betty could identify the stars she saw. Fish built a model of the stars in the neighborhood of the sun, and Betty eventually decided that the stars she had seen were in the direction of the double star system known as Zeta

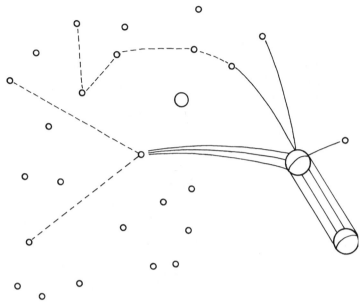

Betty Hill's star map.

Reticula; it is about 37 light-years away. Both Zeta 1 and Zeta 2, as the two stars are known, are sunlike (but slightly older than our sun); both, however, are good candidates for life. They are separated by about 400 billion miles, so planets could easily exist in ecospheres around either star. According to Betty, they were the home of the aliens.

The interpretation received considerable publicity, and it soon drew a lot of criticism. It was difficult for many to believe that she could re-draw a map accurately three years after seeing it—even if she was under hypnosis. Also, Charles Atterberg of Hanover Park, Illinois, later showed that Hill's map fit an entirely different region of the sky just as well as it fit the neighborhood of Zeta Reticula. This proved to be somewhat of an embarrassment.

The Roswell Incident

Another well-known case took place near Roswell, New Mexico, in June 1947. The fiftieth anniversary of the event was recently given enormous publicity; it made the cover of *Time* magazine. According to the story, the remains of a crashed flying saucer were found on a ranch near Roswell, with four alien bodies being found at the crash site. Several books have looked into the incident in detail (see the Bibliography). Although the books don't agree on many aspects of what happened, they all charge that there was a government cover-up.

The Roswell controversy began with the discovery by Mac Brazel of some debris on the Foster Sheep Ranch, 85 miles north of Roswell. He suggested it might be the remains of a flying saucer and turned it over to the Air Force. The Air Force later claimed it was the remains of a weather balloon. The incident got considerable publicity, but interest soon died and little was said about it until 1980 when the book *The Roswell Incident* was published by Charles Berlitz and William Moore. In the book Grady Barnett, now deceased, claims to have seen a crashed saucer about 150 miles west of the Foster Ranch in 1947; he says there were several alien bodies nearby.

Interest in the event gradually began to build, and it came to the attention of the Center for UFO Studies in Chicago in 1988. They sent out a team to investigate and three years later team members Kevin Randle and Don Schmitt published *UFO Crash at Roswell*. They charged the government had recovered both a saucer and several small alien bodies and had covered up the incident. They also mentioned an undertaker who had told them about strange activities at the military hospital in July of that year. And a nurse had told them that doctors had autopsied several aliens.

About this time Gerald Anderson, now of Springfield, Missouri, came forward claiming he had found a saucer in the desert with several dead aliens around it. According to Anderson, one was still alive. Anderson was only five at the time, but he says he vividly remembers the event. A similar tale was told by Frank Kaufman who claims he was part of a military contingent that located the wreckage of a strange object; he said he saw several small bodies inside it through a break in the side. Jim Ragsdale who now resides at Carlsbad, New Mexico, also claims to have seen a similar wreckage on the July 4 weekend of 1947.

A lot of the controversy surrounding the incident centers on a top secret document dated September 1947 that surfaced in 1984. It was delivered to the home of documentary film producer Jaime Shandera under suspicious circumstances. Shandera was meeting that day with writer Bill Moore, one of the co-authors of *The Roswell Incident*. The seven sheets of the document, which was presumably prepared for President Eisenhower, were stamped TOP SECRET; the activity the document described was

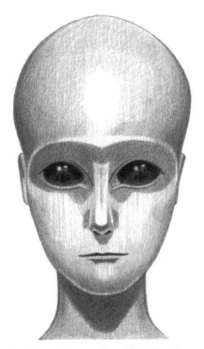

An alien of Roswell. It is similar to aliens described in stories of abductions.

referred to as Operation Majestic-12. Majestic-12 refers to the 12 members of the committee; they ranged from Truman's adviser Dr. Vannevar Bush and Admiral Roscoe H. Hillenkoetter to Dr. Detlev Bronk, chairman of the National Research Council, and aeronautical engineer Dr. Jerome Hunsaker. In the report was a discussion of the Roswell crash of 1947, and an admission that the wreckage was under study, and that four small human-like beings had been found near it. All were dead and badly decomposed, but the bodies had been examined. According to the document, news reporters and others were to be given the story that the object was a weather balloon.

The document described several examples of strange writing that had been found in the craft. Also, studies of the craft had given no clue as to how it was powered. The aliens were referred to as "Extraterrestrial Biological Entities" or EBEs. Furthermore, the document mentioned a second crash that had occurred near the Texas–Mexico border.

Is the document genuine? The circumstances surrounding its discovery are so suspicious that most people would dismiss it immediately as a fake. But surprisingly, some evidence indicates it is genuine. First of all, the Majestic team is referred to in a top secret Canadian file that has recently surfaced; there is a vague reference to it in General Nathan Twining's secret memos; and finally, there is reference to it in a briefing paper prepared for the President of the United States that was seen by Linda Moulton Howe, a Denver documentary filmmaker. Furthermore, as C.D.B. Bryan, author of *Alien Abductions, UFOs, and the Conference at MIT*, points out, the document *appears* genuine; in particular, the makeup of the Majestic-12 team is extremely likely. Still, many believe it is a fake. UFO skeptic Philip Klass is sure it is a fake; he contends that everything about the event has been blown out of proportion by the people writing books on the incident. UFOlogists, on the other hand, are not easily dissuaded; they insist there has been a cover-up of tremendous proportions by the government.

The Air Force, in its most recent report, claims the Roswell bodies were dummies that were dropped from high-altitude balloons in 1947. But most UFOlogists aren't buying the story. It's an argument that's not likely to be resolved in the near future.

Close Encounters of the Fourth Kind

As I mentioned earlier, alien abductions are now getting a lot of publicity, and surprisingly, they are also starting to be taken more seriously by scientists. Not that most of the scientists believe they are carried out by extraterrestrials; in most cases they are more interested in what is behind the phenomenon. Is it psychological? What type of people are abducted? Why are there so many reported abductions?

The conference at MIT in 1992 indicated that a surprisingly large number of people are now claiming to have been abducted. One of the speakers at the conference, Thomas E. Bullard, has catalogued 725 abduction cases. Budd Hopkins of New York has investigated 1500 cases and written three books on the phenomenon. David Jacobs, an associate professor of history at Temple University, claims to have performed 325 hypnotic sessions on more than 60 abductees. According to Jacobs, they told strikingly similar stories of being abducted by small, grayish beings with large heads. These beings were able to float them through space to an examination table within their spaceship. Almost all abductees were examined on such a table. In addition to the smaller beings, who were about four feet tall, taller ones—about six feet tall—were also sometimes present.

The "grays," as they are now commonly called, are frail and have thin limbs and no muscle or bone structure. They do not appear to have knees, elbows, ankles, or wrists, but they do have fingers—usually three or four. They have no hair on their heads or bodies. They have no ears, but they do have a nose and a mouth, with thin lips. One of their most striking features is their large black eyes, which have no pupil, iris, or cornea, and do not blink. The large aliens are Nordic in appearance; they are nonmenacing, youthful, and usually have shoulder-length blond hair.

A large number of people talked about their experiences at the conference, going into considerable detail about what happened to them during their abductions. In one case a woman on the twelfth story of a Manhattan apartment building claims she was abducted right out of her bedroom—floated to a spaceship outside her window. Strangely, her story was collaborated by two security men who were driving near her apartment. They claim they saw her and two aliens floating toward a UFO that was sitting above the building. In another case, which came to light after hypnosis, two women saw a UFO in the distance; they watched as it approached their car and hovered overhead. Two beings then appeared and the women were floated to the UFO. Both were taken into an operating room and examined. Later they were floated down to their car and released.

Despite the large number of bizarre cases presented, there was no consensus as to what was going on. Keith Thompson, generally acknowledged as one of the more scholarly of the journalists, summed up the problem: "The central paradox of the human–alien interaction is the continuing unsolvability of the UFO phenomenon by conventional means and models, coupled with the continuing manifestation of the phenomenon in increasingly bizarre forms."

A Consensus

The major question is, of course: Are the UFOs and their occupants extraterrestrial? Most astronomers and other scientists remain skeptical.

There is little doubt that a large fraction of the distant sightings of UFOs are due to natural phenomena. In general, this is assumed to be close to 90 percent of the cases reported. The most common of these natural phenomena are aircraft. Roughly 25 percent of all UFO reports are eventually shown to be due to aircraft lights. These lights can appear particularly eerie if they are seen through light clouds or fog. Planets and bright stars are also frequently mistaken for UFOs—Venus in particular, because of its brightness. About 20 percent of all reports are due to planets and bright stars.

Other objects mistaken for UFOs are fireballs (very bright meteors), weather balloons, and space satellites (the region above the Earth is littered with space vehicles and other debris). Satellites account for perhaps 5 percent of reports. Various types of electrical phenomena account for another small fraction. Plasma vortices, for example, can take on several shapes; they occur when charged particles above the Earth get caught up in a rotational motion. They have strong electromagnetic fields and could stall engines. Closely associated with them are luminous discharges of energy along fault lines. They can be caused by low-level tectonic stresses along the Earth's fault lines.

The most interesting and intriguing cases, however, are the abductees. Are the aliens real? Are they psychic? Where do they come from? According to John Mack, some of the abductees he interviewed claim that the aliens said they were from another dimension.

The French scientist Jacques Vallee has written several books on the phenomenon. He is much more broad-minded than most scientists and has seriously considered the possibility that they are extraterrestrial. In one of his books he states: "If they are an advanced race from the future, are we dealing with a parallel universe, another dimension where there are other human races living … are the UFOs 'windows' rather than objects?"

The eminent Swiss psychiatrist Carl Jung suggested in his book *Flying Saucers: A Modern Myth of Things Seen in the Sky* (1959) that they were psychic—paraphysical objects created by collective unconsciousness. He suggested that a realm of being might exist between the physical and paraphysical (it has been called the "third reality"). This third reality is presumably located between the real world of our senses and the world of abstraction or virtual reality. It is a concept that has been seriously considered by many well-known scientists.

College professor Michael Grosso suggested that they might be a "manifestation of a disturbance in the collective conscious of the human species." Another possibility is that they are "false memories." Elizabeth Loftus, a professor of psychology at the University of Washington, has

shown that false memories can be implanted—purposely and accidentally. In such cases people are convinced that an event took place—they
remember it vividly—even though it can be proven that it didn't occur.
There are, indeed, many possibilities, but the one that would be of
most interest–that the UFOs are extraterrestrial—is not one of the most
likely.

The 2001 Hypothesis
and the von Däniken Myth

If aliens have, indeed, visited us, it's much more probable that they
did so in the past. After all, we have only been around for a small fraction
of the history of the Earth—a few thousand years out of 4.6 billion. And
if aliens did visit us in the past a question that comes naturally to mind
is this: Did they leave a "calling card?"

The calling card would no doubt be some sort of artifact, which could,
of course, take many different forms. Some people have gone as far as
suggesting that aliens may have adjusted our moon in its orbit so as to
give a total eclipse. Total eclipses are, after all, exceedingly rare; they occur
only on Earth.

Artifacts were referred to as "monoliths" in the Kubrick–Clarke
movie, *2001: A Space Odyssey*. If aliens left a monolith, where would they
place it? It would obviously deteriorate rapidly on the surface of Earth. A
more likely place would be in orbit around the Earth. A satellite deriving
its energy from the sun could send out periodic radio signals. We have, of
course, seen no evidence of this, but it could be that its radio is now dead,
or the orbit it was put in was unstable, and it is now in deep space.

Another place it could be put is on the moon. If so, it could still be
there, covered with dust, waiting for us to find it.

The idea of a monolith or artifact left by ancient astronauts has been
exploited by the Swiss author Erich von Däniken. About 50 million of his
books, *Chariots of the Gods*, *Gold of the Gods*, *Gods from Outer Space*, and
so on, have been sold worldwide. Von Däniken points to many things that
he considers as evidence of ancient astronauts. He scoffs at archaeologists'
interpretations and gives his own, attributing the pyramids of Egypt to the
expertise of ancient astronauts, telling us that it would have been impossible for the Egyptians to have built them. He states that the large statues

on Easter Island in the Pacific had to be built by ancient astronauts. He reads his own interpretation into the drawings on Mayan tombs, and claims that the lines on the Nazca Plains of Peru are landing strips for astronauts (as if they would need landing strips).

Considering the sales of his books, many people are obviously taking von Däniken seriously. In general, though, his claims are unsupported and in many cases so outrageous that it is hard to believe so many could take him seriously. He is obviously playing on people's intrigue with the strange, the unknown, and the mysterious. Even though critics find most of his suggestions confusing and contradictory, he has a highly effective way of presenting them—a method that has obviously worked.

chapter 14

Epilogue

This brings us to the end of our quest for extraterrestrial life. We are still no closer to answers, but we have a better understanding of the problem, and with our new knowledge we see the universe in a different light.

We know that life began on Earth as an evolutionary process and know that it is possible that there are large numbers of extrasolar planets that likely have conditions similar to those on Earth. The process that brought us life should therefore be common. Yet the steps leading to life— particularly higher forms of life—are intensely complicated and remain beyond our realm of knowledge. So, even though astronomers are optimistic, many biologists remain pessimistic. The latter regard life on Earth as a long shot, a freak of nature. Certainly we know that if the same process—the same random assemblage of molecules—occurred on Earth again, the final product would be quite different. So if it occurred elsewhere in the universe the result is also going to be different.

Most of us have a deep-rooted hope that it *has* happened and that one day we will discover and communicate with other beings. It will be the beginning of a new day, a new sense of belonging to the universe.

But even if we detect a race of beings with our radio telescopes, it may be centuries before we are able to visit them. The tremendous difficulties of space travel, the barrier of the speed of light work against us, and unless we find a way around them, journeys through the vast, dark reaches of space will be a long time in coming.

Despite the difficulties, though, it is important that we keep searching. The rewards will far outweigh the problems. Think of it: the discovery of another race of intelligent beings. It would be the most exciting discovery ever made.

Glossary

Accretion disk: Flattened disk of matter around a star or black hole.

Adaptive optics: A technique for increasing resolution by correcting for the turbulence in the atmosphere.

Amino acid: Organic molecules that are the building blocks of protein.

Ancient astronauts: Aliens who have visited Earth in the past.

Antiparticles: Corresponding to any type of particle there is an antiparticle, usually of opposite charge. A particle and an antiparticle annihilate each other with the release of energy.

Aperiodic crystal: A crystalline structure that is not periodic.

Asteroid: Minor member of the solar system. Usually rocky. Most are between the orbits of Mars and Jupiter.

Bandwidth: Width of region in the frequency spectrum of electromagnetic waves.

Blue giants: Refers to large bright blue stars.

Bond (chemical):   A connection between two chemical elements that holds them together.

Brown dwarf: A stellar object created in the collapse of a gas cloud that is intermediate between a planet and the dimmest star.

Caldera: Collapsed region at the top of a volcano.

Carbonaceous chondrite: Dark meteorites with high carbon content.

Chaotic orbit: Orbit that is not periodic but changes chaotically.

Chemical evolution: The evolution or development of life based on chemical reactions.

Chromosome: Genetic material of cell. Contains DNA and RNA.

Coacervates: Small colloidal droplets. Possible forerunner of present cells.

Comet: A small object (few miles across) composed mostly of ice and dust. When it is close to the sun, its outer region vaporizes and it develops a tail.

Constant of the universe: Fundamental constants such as the speed of light and Planck's constant.

Constellation: An apparent group of stars.

Continental shelf: Shallow shelf around the continents.

Coronograph: Instrument for detecting the corona (outer atmosphere) of the sun.

Cosmic ray: Atomic nuclei (mostly protons) that strike the Earth's atmosphere with high speeds.

Cosmology: The study of the structure of the universe. In practice it also includes the evolution of the universe.

Cosmic background radiation: Radiation from the Big Bang explosion that created the universe. Fills the universe and has a temperature of 3 K.

Diffraction: The ability of waves to bend around corners. It is a result of the wave nature of light.

Directed panspermia: Life-containing spores purposely directed at Earth and other planets.

DNA: The master molecule of life. Contains the genetic code.

Doppler effect: The apparent change in wavelength of light due to the relative motion between a source and an observer.

Dyson sphere: Large sphere placed around a star to collect its radiant energy.

Eccentricity: A measure of the elongation or flatness of an ellipse.

Eccentric orbit: A very elongated elliptical orbit.

Ecosphere: Region around a star where water is liquid most of the time. Life zone of a star.

Electron: A negatively charged particle. One of the components of the atom.

Enzyme: An organic catalyst formed by living cells. Made up of protein.

Ethane: A colorless, odorless gas on Earth. Highly flammable. Is in liquid form on cold surfaces.

Eukaryote: Cells that contain a nucleus. A cell of considerable complexity.

Event horizon: Surface of a black hole. A one-way surface. May also exist independently of black holes.

Extrasolar: Beyond the solar system.

Fireball: Very bright meteor. Likely to leave a meteorite.

Frequency: Number of vibrations per second.

Frequency channel: A region of frequencies in the electromagnetic spectrum (usually radio region).

Galaxy: A gravitationally bound assembly of stars.

Gamma ray: Radiation of high frequency and energy. Beyond visible light.

Gene: Carries genetic material that determines physical characteristics of life.

General relativity: A theory of gravity devised by Albert Einstein.

Geothermal pools: Pools of water that are heated by the natural heat deep in the Earth.

Geothermal power: Power generated from heat deep in the Earth.

g-forces: Forces due to acceleration. Artificial gravity caused by acceleration.

Globular cluster: A roughly spherical collection of hundreds of thousands of stars.

Greenhouse effect: The partial trapping of solar radiation, similar to what happens in a greenhouse.

Heat of vaporization: Amount of heat needed to vaporize a material.

HR diagram: A plot of absolute brightness versus surface temperature for a collection of stars.

Hydrogen bonding: A relatively weak bond between hydrogen and other atoms such as oxygen or nitrogen.

Hyperspace: A higher dimensional space. Beyond our three dimensions of space.

Infrared: Region of the electromagnetic spectrum with slightly longer wavelengths than visible light.

Infrared source: A source in the sky that gives off mostly infrared radiation.

Interferometer: Refers to a collection of two or more telescopes working together as a team.

Interstellar space: Region between the stars.

Isotope: Nuclei containing same number of protons but different number of neutrons. Most elements exist in several isotopic forms.

Isotropic: The same in all directions.

Libration point: A point where the gravitational pull from two different objects is equal and opposite.

Light-year: A measure of distance. The distance a light beam travels in a year.

Luminosity: Brightness.

Macromolecules: Large chain-molecules such as DNA or RNA.

Magma: Hot molten rock.

Magnetometer: Device for measuring magnetic fields.

Main sequence: The diagonal that contains most stars in the HR diagram. Giant and dwarf stars are not in the main sequence.

Mass ratio: The ratio of the amount of fuel needed to get to a destination (usually a star) to the payload (mass at destination).

Meteorite: A rock (or other material) that survives the flight from space down through our atmosphere to the surface of Earth.

Microwaves: A region of the electromagnetic spectrum. Has longer wavelength than light.

Monolith: An artifact left by a visitor to Earth or solar system in the past.

Neuron: Basic component of the brain.

Nucleic acid: DNA and RNA.

Nuclear fission: The breaking apart of nuclei with the release of energy.

Nuclear fusion: The fusing of two nuclei. Energy is released.

Nuclear reaction: A reaction between nuclei that occurs when particles collide or decay.

Nucleotide: A unit made up of a sugar, a base, and a phosphate.

Objective: The large lens (or mirror) of a telescope.

Opposition: Two planets are in opposition when they are opposite one another in their respective orbits. They are closest at this point.

Organic molecules: Molecules of life. On Earth they are based on carbon.

Ozone: A molecule made up of three oxygens (O_3). Most oxygen is (O_2).

Panspermia: The hypothesis that life came to Earth from space via a "seed" or spore.

Parallax: The apparent shifting of a relatively nearby object with respect to a distant background of stars.

Period (of variable star): Time to go through complete cycle of brightness. Time to get back to the original brightness.

Photometer: A device for measuring the amount of light from a source.

Photosynthesis: The natural process by means of which carbon dioxide and water are converted to carbohydrates in growing plants with the aid of sunlight.

Pictogram: A "picture" that gives information.

Planck's constant: The fundamental constant of quantum theory.

Planetesimals: Rocky objects in the early universe. The building blocks of planets.

Plate tectonics: The motion of the Earth's "plates" relative to one another. Also known as continental drift.

Primitive atmosphere: Early atmosphere of Earth. A reducing atmosphere.

Prokaryote: Simple cells without nucleus, but contain DNA.

Protein: A molecular chain made up of amino acids.

Protoplanet: Early form of a planet. Forerunner of planets of today.

Purine: Made up of five-membered ring built on the side of a six-sided molecule. A and G are purines.

Pyrimidine: Consists only of six-membered rings. C and T are pyrimidines.

Quantum mechanics: Theory of atoms and molecules, their structure and interaction with radiation.

Red dwarf star: Small red star at the bottom of the main sequence.

Reducing atmosphere: Conditions in atmosphere where excess hydrogen is present.

Reflecting telescope: Telescope that uses mirror as objective.

Refracting telescope: Telescope that uses lens as objective.

Retrorocket: Rocket that is fired so that thrust is opposite the direction of travel. Slows spaceship.

RNA: "Helper" molecule in the cell. Responsible for many functions.

Scintillation: Fluctuation or change in brightness.

Seeing: Refers to the "steadiness" or stability of the atmosphere.

Singularity: A point of infinite density. At center of black hole.

Solar nebula: The whirling gas cloud that gave rise to the solar system.

Spectrum: A series of bright or dark lines that are seen when the light from a luminous object is passed through a spectroscope.

Spectral type: A classification of spectra ranging from hot stars of type O to cool stars of type N.

Spontaneous generation: The generation of life from nonlife—from such things as rotting meat.

String theory: A modern theory based on the idea that particles are made up of tiny strings.

Tachyons: Particles that travel only at speeds greater than that of light.

Telepathic: Action of one mind on another at a distance without speech. A form of communication.

Terrestrial planet: Earthlike planet.

Theory of everything: A theory that explains all phenomena in the universe.

Tidal forces: Stretching forces caused by differences in gravitational pull.

Ultraviolet radiation: A region of the electromagnetic spectrum with slightly shorter wavelengths than visible light.

Water hole: A region of the electromagnetic spectrum that is relatively free of noise.

Warp speeds: Speeds greater than the speed of light.

Wave of colonization: A "wave" that presumably should pass through our galaxy if there are many advanced civilizations.

X-ray diffraction: The bending of a beam of X rays as it passes near an opaque object or through a slit.

Bibliography

The following is a list of general and technical references for the reader who wishes to learn more about the subject. References marked with an asterisk are of a more technical nature.

Chapter 1

McDonough, T. R., *The Search for Extraterrestrial Intelligence* (New York: Wiley, 1987).
Parker, B., "Are We the Only Intelligent Life in Our Galaxy?" *Astronomy* (January, 1979), 6.
Rood, R. T., and Trefil, J., *Are We Alone?* (New York: Scribner, 1981).

Chapter 2

Asimov, I., *The Genetic Code* (New York: Orion, 1962).
Gribbin, J., *In Search of the Double Helix* (New York: McGraw-Hill, 1985).
Judson, H. F., *The Eighth Day of Creation* (London: Jonathan Cape, 1979).
Ponnamperuma, C., *The Origin of Life* (New York: NAL/Dutton, 1972).
Schrödinger, E., *What Is Life?* (London: Cambridge University Press, 1967).
Watson, J., *The Double Helix* (New York: NAL/Dutton, 1968).

Chapter 3

Asimov, I., *Beginnings: The Story of the Origin—Of Mankind, Life, the Earth, the Universe* (New York: Walker, 1987).

Breuer, R., *Contact with the Stars* (New York: Freeman, 1982).

Cairns-Smith, A. G., *Seven Clues to the Origin of Life* (London: Cambridge University Press, 1985).

*Calvin, M., *Chemical Evolution* (New York: Oxford, 1969).

Dyson, F., *Origin of Life* (London: Cambridge University Press, 1985).

*Fox, S., and Dose, K., *Molecular Evolution and the Origin of Life* (New York: Freeman, 1972).

Goldsmith, D., and Owens, T., *The Search for Life in the Universe* (Reading: Addison-Wesley, 1972).

Gribbins, Jr., *Genesis* (New York: Delacorte, 1981).

Ponnamperuma, C., *The Origin of Life* (New York: NAL/Dutton, 1972).

Chapter 4

Anderson, P., *Is There Life on Other Worlds?* (New York: Collier Books, 1963).

Fredrickson, J., and Onstott, T., "Microbes Deep Inside the Earth," *Scientific American* (October, 1976) 68.

Gunn, J., *Alternate Worlds* (Englewood Cliffs: Prentice-Hall, 1975).

McDonough, T. R., *The Search for Extraterrestrial Intelligence* (New York: Wiley, 1987).

Nichols, P. (editor), *The Science Fiction Encyclopedia* (New York: Dolphin Books, 1979).

Ponnamperuma, C., and Cameron, A., *Interstellar Communication: Scientific Perspective* (Boston: Houghton Mifflin, 1974).

Chapter 5

Horowitz, N., *To Utopia and Back: The Search for Life in the Solar System* (New York: Freeman, 1986).

Hoyt, W. G., *Lowell and Mars* (Tucson: University of Arizona Press, 1976).

Kargel, J., and Strom, R., "Global Climatic Changes on Mars," *Scientific American* (November, 1996) 80.

McKay, C., "Did Mars Once Have Martians?" *Astronomy* (September, 1993) 26.

Ridpath, I., *Messages from the Stars* (New York: Harper-Collins, 1978).

Sheehan, W., *The Planet Mars* (Tucson: University of Arizona Press, 1997).

Shklovskii, I. S., and Sagan, C., *Intelligent Life in the Universe* (New York: Dell, 1966).

Vaucouleurs, G. de, *The Planet Mars* (London: Faber and Faber, 1950).

Chapter 6

Beatty, J., "Life from Ancient Mars," *Sky and Telescope* (October, 1996) 18.
Bennington, D., "Looking for Life," *Stardate* (November/December, 1996) 4.
Goldsmith, D., *The Hunt for Life on Mars* (New York: NAL/Dutton, 1997).
McKay, C., "Looking for Life on Mars," *Astronomy* (August, 1997) 38.
*McKay, D., et al., "Search for Past Life on Mars: Possible Relic Biogenic Activity in Martian Meteorite ALH84001," *Science* (August, 1996) 924.
Naeye, R., "Was There Life on Mars?" *Astronomy* (November, 1996) 38.
Swartz, M., "It Came from Outer Space," *Readers Digest* (March, 1997) 100.

Chapter 7

Benningfield, D., "A Better, Faster, Cheaper, Route to Mars," *Stardate* (November/December, 1996) 16.
Cray, D., Downer, J., and Thompson, D., "Uncovering the Secrets of Mars," *Time* (July, 1997) 26.
Shibley, J., "Back to Mars on All Sixes," *Astronomy* (January, 1997) 48.
Zubrin, R., *The Case for Mars* (New York: Free Press, 1996).

Chapter 8

Burnham, R., "Hubble Maps Titan's Hidden Landscape," *Astronomy* (February, 1995) 44.
Hoagland, R., "The Europa Enigma," *Star and Sky* (January, 1980) 16.
Milstein, M., "Diving into Europa's Ocean," *Astronomy* (October, 1997) 38.

Chapter 9

Black, D., "Other Suns, Other Planets," *Sky and Telescope* (August, 1996) 20.
Black, H., and Monroe, S., "Is There Life in Outer Space?" *Time* (February, 1996) 50.
Dole, S., *Habitable Planets for Man* (New York: Elsevier, 1964).
Goldsmith, D., *Worlds Unnumbered* (Sausalito: University Science Books, 1997).
Halpern, P., *The Quest for Alien Planets* (New York: Plenum, 1997).
Mammana, D., and McCarthy, D., Jr., *Other Suns, Other Worlds* (New York: St. Martin's Press, 1995).
Naeye, R., "Is This Planet for Real?" *Astronomy* (March, 1996) 34.
Naeye, R., "Two New Solar Systems," *Astronomy* (April, 1996) 50.
Naeye, R., "The Strange New Planetary Zoo," *Astronomy* (April, 1997) 42.

Roth, J., and Sinnott, R., "Our Nearest Celestial Neighbors," *Sky and Telescope* (October, 1996) 32.

Chapter 10

Bracewell, R., *The Galactic Club* (New York: Freeman, 1975).

Breuer, R., *Contact with the Stars* (New York: Freeman, 1982).

Christian, J. (editor), *Extraterrestrial Intelligence: The First Encounter* (Amherst: Prometheus Books, 1976).

Drake, F., and Sobel, D., *Is Anyone Out There?* (New York: Delacorte, 1992).

Goldsmith, D., and Owen, T., *The Search for Life in the Universe* (Reading: Addison-Wesley, 1992).

McDonough, T. R., *The Search for Extraterrestrial Intelligence* (New York: Wiley, 1987).

Ponnamperuma, C., and Cameron, A., *Interstellar Communication: Scientific Perspective* (Boston: Houghton Mifflin, 1974).

Shotak, S., "When ET Calls Us," *Astronomy* (September, 1997) 37.

Tarter, J., "Searching for THEM: Interstellar Communications," *Astronomy* (October, 1982) 6.

Chapter 11

Davies, P., *Are We Alone?* (New York: Basic Books, 1995).

Papagiannis, M., "The Search for Extraterrestrial Civilizations: A New Approach," *Mercury* (January/February, 1982) 12.

Papagiannis, M., "Bioastronomy: The Search for Extraterrestrial Life," *Sky and Telescope* (June, 1984) 508.

Rood, R., and Trefil, J., *Are We Alone?* (New York: Scribner, 1981).

Tipler, F., "The Most Advanced Civilization in the Galaxy is Ours," *Mercury* (January/February, 1982) 5.

Chapter 12

Forward, R., and Davis, J., *Mirror Matter: Pioneering Antimatter Physics* (New York: Wiley, 1988).

Friedman, L., *Starsailing: Solar Sails and Interstellar Travel* (New York: Wiley, 1988).

Macvey, J., *Interstellar Travel: Past, Present, and Future* (New York: Stein and Day, 1977).

Oberg, J., and Oberg, A., *Pioneering Space: Living in the Next Frontier* (New York: McGraw-Hill, 1986).

O'Neill, G., *The High Frontier: Human Colonies in Space* (New York: Morrison, 1977).

Chapter 13

Barclay, D., and Barclay, T. (editors), *UFOs: The Final Answer* (Lanham: Barnes & Noble Books, 1993).

Blum, H., *Out There* (New York: Pocket Books, 1990).

Bryan, C. D. B., *Alien Abductions, UFOs, and the Conference at MIT* (New York: Knopf, 1995).

Condon, E. U., *Scientific Study of UFOs* (New York: Bantam, 1969).

David, J. (editor), *The Flying Saucer Reader* (New York: NAL/Dutton, 1967).

Fuller, J. G., *The Interrupted Journey* (New York: Dial, 1966).

Hynek, J., *The UFO Experience: A Scientific Enquiry* (New York: Ballantine, 1974).

Hynek, J., and Vallee, J., *The Edge of Reality* (Washington: Regnery, 1975).

Jaroff, L., "Did Aliens Really Land?" *Time* (June, 1997) 68.

Jung, C. G., *Flying Saucers: A Modern Myth of Things Seen in the Sky* (New York: NAL/Dutton, 1959).

Klass, P. J., *UFO Abductions: A Dangerous Game* (Amherst: Prometheus Books, 1989).

Mack, J., *Abductions: Human Encounters with Aliens* (New York: Scribner, 1994).

Index